Analysis of Longitudinal Data with Examples

Analysis of Longitudinal Data with Examples

You-Gan Wang
Liya Fu
Sudhir Paul

CRC Press
Taylor & Francis Group
Boca Raton London New York

CRC Press is an imprint of the
Taylor & Francis Group, an **informa** business

A CHAPMAN & HALL BOOK

First edition published 2022
by CRC Press
6000 Broken Sound Parkway NW, Suite 300, Boca Raton, FL 33487-2742

and by CRC Press
2 Park Square, Milton Park, Abingdon, Oxon, OX14 4RN

CRC Press is an imprint of Taylor & Francis Group, LLC

ISBN: 978-1-4987-6460-5 (hbk)
ISBN: 978-1-032-19652-7 (pbk)
ISBN: 978-1-315-15363-6 (ebk)

DOI: 10.1201/9781315153636

Publisher's note: This book has been prepared from camera-ready copy provided by the authors.

To those who have passions in longitudinal data analysis

Contents

List of Figures

List of Tables

Preface

Longitudinal data are ubiquitous in economics, medical studies, and environmental research. The fundamental framework in statistics is regression, where many problems and solutions can be embedded. Generally, the solution hinges on a linear combination of the data for predictors, parameter estimators, or the estimating functions. In a regression framework, each observation is regarded as a "replicate" as the difference is taken care of by the nonzero coefficients of the regressors.

Why do we need to model the correlations? A simple example is the paired t-test scenario where the paired data are correlated. Ignoring such correlation will lead to misuse of the independent t-test, which will produce misleading inferences. Analysis of correlated data needs to account for the correlations of each observation with all the rest observations. This defining correlation feature of longitudinal data makes modeling dependence a key topic in longitudinal data analysis. The dependence structure will imply how we borrow the information from each other for better prediction. Application of which or what skill depends on how you describe or model the underlying correlations. Random effects models are the most intuitive as an extension of linear regression, or generalized linear models, while marginal models choose to model the correlation structures directly. The concept of correlation herein is the same as in time series models. The unique issue is that the number of observations for each subject is usually small, and the inference will need to rely on independent "replicates" from many subjects.

In fact, correlation must be accounted for *properly* in order to obtain valid inferences. For example, in longitudinal studies, the sequential nature of the measures implies that certain types of correlation structures are likely to arise. Therefore, careful modeling is needed to make the hypothesized model as close as possible to the true one. Chapter 2 presents eight examples, and all the datasets are available for readers to investigate further.

Chapter 3 introduces the basic components of the statistical models for longitudinal data analysis. While the correlation modeling is the key, such model and estimation become meaningless when the other parts are not modeled properly.

All statistical models have explicit and implicit assumptions to work. This does not necessarily mean the model does not work if certain assumptions are in doubt or violated. It is, therefore, of great both practical and theoretical interest to investigate misspecification implications. This is why the GEE (Generalized Estimating Equation) approach is elegant; the correlation structure is just a "working" correlation model, and consistency is not affected when it is incorrectly specified, but the framework allows you to reach optimality if the specified correlation model is very close to the true one. The likelihood-based estimation is important as it lays the foundation for estimation and inference. If we are interested in quantifying the uncertainties in the resultant estimates (such as standard errors and confidence intervals), we would have to investigate implications of any misspecified model components, *i.e.*, the variance and correlation parts in our cases. Chapter 4 presents various parameter estimation approaches for the mean and variance functions and correlation structures.

The marginal model consists of three components, the mean function, variance function, and correlation structure. Chapter 5 introduces the criteria for selecting each of these three key parts. Selecting the useful predictors (regressors) in the mean function is a classical topic. We have only introduced quasi-likelihood and Gaussian likelihood method in this respect for longitudinal data analysis. In fact, the Lasso approach using L_1 norm is also applicable, although it is often used when the number of predictors is very high.

In longitudinal studies, the collected data often deviates from normality, and the response variable and/or predictors may contain some potential outliers, which often results in serious problems for variable selection and parameter estimation. Therefore, robust methods have attracted much attention in recent years. Robust approaches using rank and quantile regression are given in Chapter 6.

Clustered data refer to a set of measurements collected from subjects that are structured in clusters, which arise in many biostatistical practices and environmental studies. Responses from a group of genetically-related members from a familial pedigree constitute clustered data in which the responses are correlated. Chapter 7 is devoted to the methodology of such data analysis.

Missing data or missing values occur when no information is available on the response or some of the predictors or both the response and some predictors for some subjects who participate in a study of interest. There can be a variety of reasons for the occurrence of missing values. Nonresponse occurs when the respondent does not respond to certain questions due to stress, fatigue, or lack of knowledge. Some individuals in the study may not respond because some questions are sensitive. Missing data can create difficulty in the analysis because nearly all standard statistical methods presume complete information for all the variables included in the analysis. Chapter 8 deals with methodologies for the analysis of longitudinal data with missing values.

The traditional designed experiments involve data that need to be analyzed using a fixed-effects model or a random-effects model. Central to the idea of variance components models is the idea of fixed and random effects. Each effect in a variance components model must be classified as either a fixed or a random effect. Fixed effects arise when the levels of an effect constitute the entire population about which one is interested. Chapter 9 develops the methodology of analyzing longitudinal data using random effects and transitional models.

High-dimensional longitudinal data involving many variables are often collected. The inclusion of redundant variables may reduce accuracy and efficiency for both parameter estimation and statistical inference. When a large number of predictors are collected as in phenotypical studies to identify responsible genes, the inclusion of redundant variables can reduce the accuracy and efficiency of estimation and prediction (inflated false discovery rate and reduced power). Therefore, it is important to select the appropriate covariates in analyzing longitudinal data. However, it is a challenge to select significant variables in longitudinal data due to underlying correlations and unavailable likelihood. Chapter 10 presents how the lasso-type approach can be used in longitudinal data analysis.

This book is written for (1) applied statisticians and scientists who are interested in how dependence is taken care of in analyzing longitudinal data; (2) data analysts who are interested in the development of techniques for longitudinal data analysis; and (3) graduate students and researchers who are interested in researching on correlated data analysis. This book is also suitable as a graduate textbook to assist the students in learning advanced statistical thinking and seeking potential projects in correlated data analysis.

Author Bios

Professor Wang obtained his Ph.D. in dynamic optimization in 1991 (University of Oxford) and worked for CSIRO (2005–2010). Before returning to Australia, Professor Wang worked for the National University of Singapore (2001–2005) and Harvard University as Assistant Professor and Associate Professor (1998–2000) in biostatistics. He joined the University of Queensland in April 2010 as Chair Professor of Applied Statistics to lead the Centre for Applications in Natural Resource Mathematics and to promote applied statistics and mathematics. Currently, he is Capacity Building Professor in Data Science at Queensland University of Technology, Australia.

Professor Wang has developed a number of novel statistical methodologies in longitudinal data analysis published by top statistical journals (*Biometrika, Biometrics, Statistics in Medicine, Journal of the American Statistician Association, Annals of Statistics*). His recent interests and successes include (1) "working" likelihood approach for hyperparameter estimation and model selection, (2) integrating statistical learning and machine learning for dependent data analysis, and (3) data-driven approach for robust estimation. More recently, he advocates "working" likelihood approaches to parameter estimation but recognizing possibly a different likelihood that generating the observed data in inferencing. This has been found very useful in finding data-dependent tuning parameters in robust estimation and hyper-parameters in machine learning algorithms. He has published over 175 papers in international SCI journals.

Liya Fu obtained her Ph.D. in 2010 from Northeast Normal University. Currently, she is an Associate Professor of Statistics at Xi'an Jiaotong University. She worked briefly as a Postdoctoral Fellow at the University of Queensland, after two-years, visiting students at CSIRO (2008–2010), Australia. Dr. Fu mainly focuses on the methodologies for the analysis of longitudinal data and has published about more than 20 papers in international journals, including *Biometrics*, *Statistics in Medicine*, *Journal of Multivariate Analysis*.

Professor Sudhir Paul obtained his Ph.D. in 1976 (University of Wales). He worked as a postdoctoral fellow (University of Newcastle Upon Tyne, 1976–1978) and a lecturer (University of Kent at Canterbury, 1978-1982) before moving to Canada in 1982. He started as Assistant Professor at the University of Windsor and moved through all professorial ranks and finally, in 2005, became a distinguished University Professor. He became a fellow of the Royal Statistical Society in 1982 and a fellow of the American Statistical Association in 1986.

Professor Paul has developed many methodologies for the analyses of over-dispersed and zero-inflated count data, longitudinal data, and familial data and published in most of the top-tier journals in statistics (*Journal of the Royal Statistical Society, Biometrika, Biometrics, Journal of the American Statistician Association, Technometrics*). Professor Paul supervised over 50 graduate students including 16 Ph.D. students, and has published over 100 papers.

Contributors

You-Gan Wang
School of Mathematical
   Sciences,
Queensland University of
   Technology,
Brisbane, Australia

Liya Fu
School of Mathematics &
   Statistics,
Xi'an Jiaotong University,
Xi'an, China

Sudhir Paul
Department of
   Mathematics &
   Statistics
University of Windsor,
Ontario, Canada

Acknowledgment

This book is partially supported by Australian Research Council (ARC) Discovery Project (DP160104292), the Australian Research Council Centre of Excellence for Mathematical and Statistical Frontiers (ACEMS), under grant number CE140100049, and the Natural Science Foundation of China (No. 11871390). We also wish to thank Dr. Qibin Duan, Dr. Jiyuan An, Mr. Ryan (Jinran) Wu and Miss Jiaqi Li for some assistance in latex work.

Chapter 1

Introduction

1.1 Longitudinal Studies

Understanding the variance and covariance structure in the data is an imperative part in statistical inferences.

Quantification of parameter values and their standard errors critically depends on the underlying structure that generated the data. The challenge of a modeler is **to specify** what the average or mean function should be and how we describe the most appropriately what the variance/covariance structure in the data. Such mean functions will be crucial in forecasting, and the variance/covariance functions are often used as weighting and to describe the uncertainties in future observations.

Longitudinal data are routinely collected in this fashion in a broad range of applications, including agriculture and the life sciences, medical and public health research, and industrial applications. For longitudinal studies, the same experimental units (such as patients, trees, and sites) are observed or measured for multiple times over a period of time. The experiment units can be patients in medical studies, trees in forestry studies, and animals in biological studies. The experimental units can also be sites or buildings where water or air quality data are collected. This repeating nature or clustering nature exhibited in the data makes the longitudinal data behold certain variance/covariance structures that we need to reflect in our model.

Longitudinal data contains temporal changes over time from each individual. In contrast to a cross-sectional study in which a single outcome is measured for each individual, the prime advantage of a longitudinal study is its effectiveness for studying changes over time. Therefore, observations from the same patients are correlated, and this correlation must be taken into account in statistical analysis. Thus, it is necessary for a statistical model to reflect the way in which the data were collected in order to address these questions.

We assume that the response variable from subject i at time j is represented as y_{ij}, which has mean μ_{ij} and variance $\phi\sigma_{ij}^2$ for $j = 1, \ldots, n_i$ and $i = 1, \ldots, N$. Here ϕ is an unknown scale parameter, μ_{ij} and σ_{ij}^2 are some known functions **with unknown parameters** of the covariates, and X_{ij} is a $p \times 1$ vector. Let $\mu_i = (\mu_{ij})$ be the marginal mean vector for subject i. Let $Y_i^{\mathrm{T}} = (y_{i1}, \ldots, y_{in_i})$. The covariance matrix of Y_i, V_i, is assumed to have the form $\phi A_i^{1/2} R_i(\alpha) A_i^{1/2}$, in which $A_i = \mathrm{diag}(\sigma_{ij}^2)$ and $R_i(\alpha)$ is the correlation matrix parameterized by α, a q-vector.

DOI: 10.1201/9781315153636-1

We first consider the special case when $n_i = 1$ and all observations are independent of each other. This is basically the setup of the generalized linear models (GLMs). In GLMs, we have $\mu_{ij} = h(X_{ij}^T\beta)$, a function of some linear combination of the covariates, and the variance σ_{ij}^2 is related to the covariates via some *known* functions of the mean.

Later on we will allow multiple observations from the same subjects, and correlations will have to be allowed among these within-subject observations while observations from different subjects are assumed independent.

In longitudinal studies, a variety of models can be used to meet different purposes of the research. For example, some experiments focus on individual responses; the others emphasize the average characters. Two popular approaches have been developed to accommodate different scientific objectives: the random effects model and the marginal model (Liang et al., 1992).

The random-effects model is a subject-specific model that models the source of the heterogeneity explicitly. The basic premise behind the random-effects model is that we assume that there is natural heterogeneity across individuals in a subset of the regression coefficients. That is, a subset of the regression coefficients is assumed to vary across individuals according to some distribution. Thus the coefficients have an interpretation for individuals.

Marginal model is a population-average model. When inferences about the population-average are the focus, the marginal models are appropriate. For example, in a clinical trial, the average difference between control and treatment is most important, not the difference for a particular individual.

The main difference between the marginal and random-effects model is the way in which the multivariate distribution of responses is specified. In a marginal model, the mean response modeled is conditioning on only fixed covariates, while for random-effects models, it is conditioned on both covariates and random effects.

The random-effects models can be seen as a likelihood-based approach, while the marginal approach is semiparametric in the sense that only the mean function and the **covariance** structure are modeled via some parametric functions (and likelihood function is avoided).

1.2   Notation

In general, we will use capital letters to represent vectors and small letters to present scalars. The letters represent random variables or observations depending on the context to distinguish.

N: the number of subjects.
(y_{ij}, X_{ij}, t_{ij}): data from subject i, $1 \le j \le n_i$ and $i = 1, 2, \ldots, N$.
y_{ij}: responses from subject i at time t_j.
μ_{ij}: mean of y_{ij}.
$\text{var}(y_{ik}) = \sigma_{ij}^2 = \phi v(\mu_{ij})$.
$g(\cdot)$: the link function.
$\beta = (\beta_1, \ldots, \beta_p)^T$: parameter vector of dimension p in the mean function.

$\tilde{\beta}$: the true value of β.

α: parameter vector in the correction matrix.

γ: parameter vector in the variance-mean matrix.

$$Y_i = \begin{pmatrix} y_{i1} \\ \vdots \\ y_{in_i} \end{pmatrix}, \quad Y = \begin{pmatrix} Y_1 \\ \vdots \\ Y_N \end{pmatrix}, \quad X_i = \begin{pmatrix} X_{i1}^{\mathrm{T}} \\ \vdots \\ X_{in_i}^{\mathrm{T}} \end{pmatrix} = \begin{pmatrix} x_{i11} & \cdots & x_{i1p} \\ x_{i21} & \cdots & x_{i2p} \\ \vdots & \vdots & \vdots \\ x_{in_i1} & \cdots & x_{in_ip} \end{pmatrix},$$

in which $X_{ij} = (x_{ij1}, \ldots, x_{ijp})^{\mathrm{T}}$, $j = 1, 2, \ldots, n_i$.

$$\mu_i = \mathbf{E}(Y_i) = \begin{pmatrix} \mu_{i1} \\ \vdots \\ \mu_{in_i} \end{pmatrix}, \quad V_i = \mathrm{var}(Y_i) = \begin{pmatrix} \mathrm{var}(Y_{i1}) & \cdots & \mathrm{Cov}(Y_{i1}, Y_{in_i}) \\ \mathrm{Cov}(Y_{i2}, Y_{i1}) & \ddots & \mathrm{Cov}(Y_{i2}, Y_{in_i}) \\ \vdots & \ddots & \vdots \\ \mathrm{Cov}(Y_{in_i}, Y_{i1}) & \cdots & \mathrm{var}(Y_{in_i}) \end{pmatrix}.$$

The true parameter vector is denoted as $\tilde{\beta}$, i.e., we can decompose Y_i as

$$Y_i = g(X_i\tilde{\beta}) + \epsilon_i,$$

in which $\epsilon_i = (\epsilon_{i1}, \ldots, \epsilon_{in_i})^{\mathrm{T}}$, and $\mathbf{E}(\epsilon_i) = \mathbf{0}$.

Chapter 2

Examples and Organization
of The Book

We now introduce some longitudinal studies. Some of the datasets will be used for illustration through this book.

2.1 Examples for Longitudinal Studies

2.1.1 HIV Study

In a medical study, a measure of viral load may be taken at monthly intervals from patients with HIV infection. The objective was to establish viral dynamics (Kinetics of HIV and CD4 turnover). For example, Ho et al. (1995) concluded a mean virus half-life of about 2 days and a mean production of $0.68 \pm 0.13 \times 10^9$ virion per day. Elucidation of HIV dynamics greatly advanced AIDS research since HIV was first identified in 1983.

In the AIDS Clinical Trial Group (ACTG) Protocol 315 (Wu and Ding, 1999), 53 HIV infected patients were treated with potent antiviral drugs (ritonavir monotherapy for the first 10 days, 3TC and AZT added on day 10), which consists of protease inhibitor (PI) and reverse transcriptase inhibitor (RTI) drugs. Plasma HIV-1 RNA copies was repeatedly measured on days 2, 7, 10, 14, 21, 28 and weeks 8, 12, 24, and 48 after initiation of treatment. The nucleic acid sequence-based amplification assay (NASBA) was used to measure plasma HIV-1 RNA. This assay offers a lower limit of detection of 100 RNA copies/ml plasma. Five patients discontinued the study due to drug intolerance and other problems. **Therefore, the observations were collected from only** 48 **patients.** Figure 2.1 shows the HIV viral load measurements (plasma HIV-1 RNA copies) for 48 valuable patients.

2.1.2 Progabide Study

A clinical trial was conducted in which 59 people with epilepsy suffering from simple or partial seizures were assigned at random to receive either the anti-epileptic drug progabide (subjects 29-59) or an inert substance (a placebo, subjects 1-28). Because each individual might be prone to different rates of experiencing seizures, the investigators first tried to get a sense of this by recording the number of seizures suffered

DOI: 10.1201/9781315153636-2

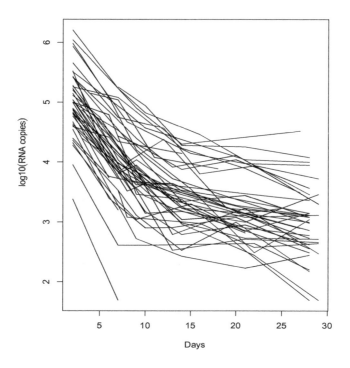

Figure 2.1 *Parallel lines of log(RNA copies) in HIV study.*

by each subject over the 8-week period prior to the start of administration of the assigned treatment. It is common in such studies to record such baseline measurements
so that the effect of treatment for each subject may be measured relative to how that
subject behaved before treatment. Following the commencement of treatment, the
number of seizures for each subject was counted for each of four 2-week consecutive periods. The age of each subject at the start of the study was also recorded, as it
was suspected that the age of the subject might be associated with the effect of the
treatment somehow. The primary objective of the study was to determine whether
progabide reduces the rate of seizures in subjects like those in the trial.

The data for the first five subjects in each treatment group are shown in Table 2.1.
The boxplots of the number of seizures for epileptics (see Figure 2.2) indicate that
there exist outliers. Thus, a robust method should be considered when analyzing this
dataset. Figure 2.3 indicates that there are also strong within-subject correlations in
the epileptic data, as reported in Thall & Vail (1990). Thall and Vail (1990) presented
a number of covariance models to account for overdispersion, heteroscedasticity, and
dependence among repeated observations.

Table 2.1 *The seizure count data for 5 subjects assigned to placebo (0) and 5 subjects assigned to progabide (1).*

	Period						
Subject	1	2	3	4	Trt	Baseline	Age
1	5	3	3	3	0	11	31
2	3	5	3	3	0	11	30
3	2	4	0	5	0	6	25
4	4	4	1	4	0	8	26
5	7	18	9	21	0	66	22
		⋮					
29	11	14	9	8	1	76	18
30	8	7	9	4	1	38	32
31	0	4	3	0	1	19	20
32	3	6	1	3	1	10	30
33	2	6	7	4	1	19	18

2.1.3 Hormone Study

In this study, a total of 492 urine samples were collected from 34 women (with menstrual cycles) aged between 27 and 45 years, and urinary progesterone was assayed on alternate days. Each woman contributed from 11 to 28 observations over a period of time; hence the data are unbalanced. One purpose of the study was to test the effect of age and body mass index (BMI) on women's progesterone level after an appropriate adjustment of their menstrual cycles. More details can be found in Sowers et al. (1998). Figure 2.4 indicates that the log-transformed progesterone level exhibits a nonlinear effect in time. Zhang et al. (1998) and Fung et al. (2002) found that some outliers exist in the data and fitted the data using a semiparametric mixed model, which included covariates, age, body mass index BMI, and nonlinear time effects. In Chapter 5, we will use a quintic polynomial to fit the time effect. In addition to age, BMI, and time effect, we will also consider their interaction effects.

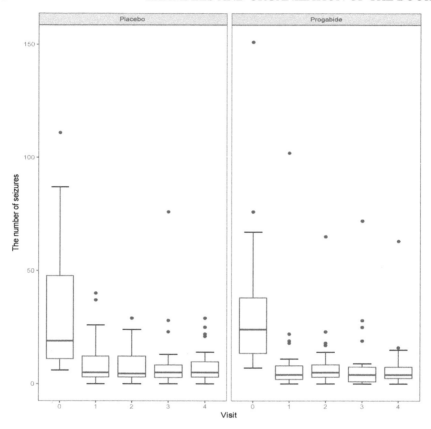

Figure 2.2 *Boxplots of the number of seizures for epileptics at baseline and four subsequent 2-week periods: "0" indicates baseline.*

2.1.4   Teratology Studies

Paul (1982) reported that the mean litter sizes from female banded Dutch rabbits are found to be $7.96, 7.00, 7.19$, and 5.94 when exposed to control, low, medium and high levels of toxic doses. The corresponding observed malformation rates are 0.135, $0.135, 0.344$ and 0.228, i.e. the normality rates are $(0.865, 0.865, 0.656, 0.772)$. The medium dose level, instead of the high dose level, appears to have the lowest survival rate (Williams, 1987). As we will see, this is probably due to ignoring the death *in utero*.

However, the average numbers of normal fetuses produced by each dam are $6.889, 6.052, 4.714$, and 4.588. The ratios to the control group are $(0.878, 0.684, 0.666)$, which exhibits a decreasing trend. The dramatic reduction in an average

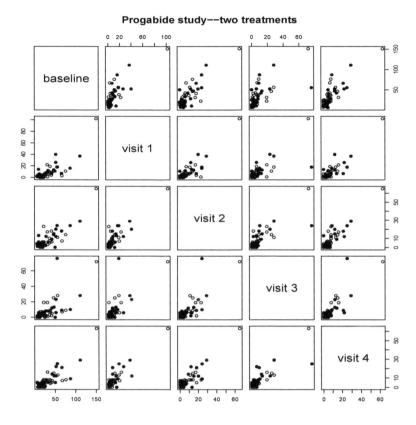

Figure 2.3 *Scatterplot matrix for epileptics at baseline and four subsequent 2-week periods: solid circle for placebo; hollow circle for progabide.*

number of normal fetuses per dam clearly shows the adverse effects of this toxic agent. By now, without any parametric modeling, we can see an obvious adverse effect from this monotonic trend. The adjusted proportions of normal fetuses for the four dose levels are (0.865, 0.760, 0.592, 0.576) assuming N, the average number of initial population size for each group, is 7.96, as observed in the control group. These proportions of normal fetuses become (0.765, 0.673, 0.524, 0.509) if we assume $N = 9$. Note that for any values assumed for N, the ratios to the control group are still (0.878, 0.684, 0.666). This shows that roughly the excessive risk for the low, medium, and high dose groups are 0.122, 0.316, and 0.334 (the actual dose values are not available).

Parametric models are usually used to quantify the relationship with dose levels and other covariates. This will allow us to determine the virtually safe dose (Ryan, 1992). Traditional risk assessment is based on the probabilities of abnormal fetuses within each litter. The risk of death *in utero* is not accounted for because the inference

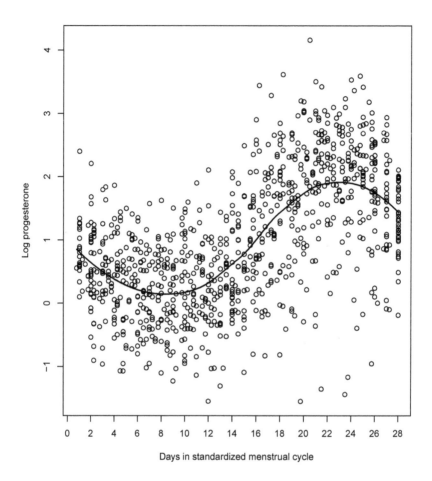

Figure 2.4 *Progesterone levels (on log scale) plot against the number of days in a standardized menstrual cycle. Solid line indicates the estimated population mean curve.*

is based on successful implantations. The additional risk of death *in utero* due to the toxic agent has to be modeled to obtain the overall risk. Estimation of the overall risk ratios for different dose levels will depend on modeling the average response (including death *in utero*) numbers per dam, which may not be directly observed or reliably recorded. If death *in utero* is ignored, agent risk will be underestimated.

Development of offspring is subject to the risk of uniforming due to toxins, although the number of such littermates may not always be fully observed. This risk can also be interpreted as unsuccessful implantation. The offspring will be further subject to adverse effects after successful implantation. It seems appropriate to take into account of death *in utero* as well as other observed abnormalities in fetuses in risk assessment, although this is not currently required by regulatory agencies such as Environmental Protection Agency and Food and Drug Administration.

2.1.5 Schizophrenia Study

We use the data set from the Madras Longitudinal Schizophrenia Study available from Diggle et al. (2002) as an example for binary data. This study investigated the course of positive and negative psychiatric symptoms over the first year after initial hospitalization for schizophrenia. The response variable (Y) is binary, indicating the presence of thought disorder. The covariates are (i). Month - duration since hospitalization (in months), (ii). Age - age of patient at the onset of symptom (1 represents Age < 20; 0 otherwise), (iii). Gender - gender of patient (1=female; 0=male), (iv). Month×Age - Interaction term between variables Month and Age, and (v). Month×Gender - Interaction term between variables Month and Gender.

2.1.6 Labor Pain Study

In this study, 83 women were randomized to either the placebo group (40 women) or the active treatment group (43 women). Treatment was initiated when cervical dilation was 8cm. The self-reported amount of pain was recorded on a 100mm line (0=no pain, 100=very much pain) in 30-minute intervals over a period of 180 minutes. The length of the line to the left of the subject's perceived pain level was measured to the nearest 0.5 mm; therefore, the outcome variable is essentially continuous. In the latter part of the study, there were many missing values. We assume the data are missing completely at random. More details can be found in Davis (1991).

Parallel lines and boxplots (Figures 2.5 and 2.6) indicate that pain scores seem to be higher in the placebo group than in the active group at each point. The distribution of the pain score is extremely non-normal (see Figure 2.7 for the placebo group and active treatment group, respectively); thus, the use of nonparametric method is indicated.

2.1.7 Labor Market Experience

We now consider a survey example, a study of the National Longitudinal Survey of Labor Market Experience (Center for Human Resource Research 1989). Years of education, age, income in each year, and location of the subjects were collected. A subsample of 3,913 women between 14 and 26 years old who have completed their education and earning wages in excess of $1/hour but less than $700/hour are used in the analysis. The incomes for each woman were collected.

The response variable in this example is the annualized income (assuming 2000 working hours a year). The log-transformed annualized incomes for the cohort is analyzed first, followed by a re-analysis using the untransformed incomes. We are interested in how education affects the pay rate and if there is a difference between southerners and others. Figures 2.8 and 2.9 indicate that the increase of average

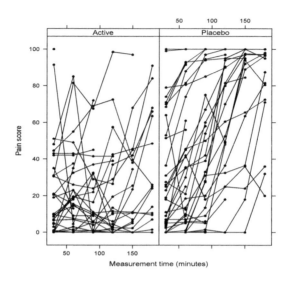

Figure 2.5 *Parallel lines of pain scores for active and placebo groups.*

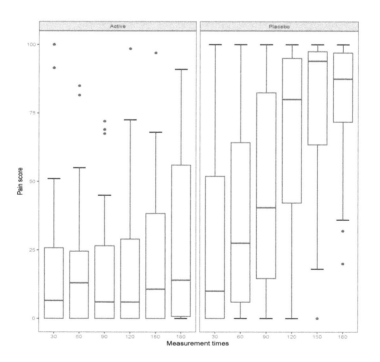

Figure 2.6 *Boxplots of pain scores for placebo and active groups.*

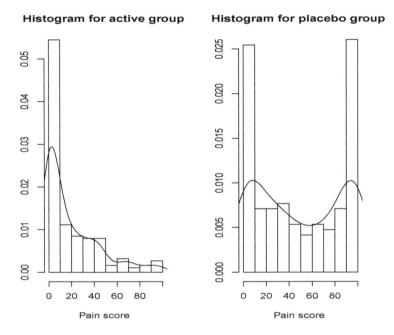

Figure 2.7 *Histograms of pain scores for active and placebo groups.*

incomes as education time increases, and the average income of south is lower than that in others when women have the same education.

When modeling this dataset, the candidate covariates are (i) `age`: the age at the time of the survey; (ii) `grade`: the current grade that has been completed; and (iii) `south`: whether a person came from the south.

2.1.8   Water Quality Data

This dataset was collected from 32 sites from 1997 to 2007 at a dam in South East Queensland, Australia. There are 12 water quality indicators monitored, including Total Cyanophytes, Chlorophyll A, and Ammonia Nitrogen. An important water quality indicator is the total amount of cyanophytes in the water because almost all toxic freshwater blooms are caused by cyanophytes. Cyanophyte blooms form unpleasant surface scums and cause odor in the water. Some cyanophyte is toxic and can cause deaths. Cyanophyte is known to be affected by season and the rate of flow of water. We present repeated measurements of the values of total cyanophyte from four sites in Figure 2.10. Figure 2.10 indicates that there exist some outliers. The Q-Q plot shows that the distribution of the total cyanophyte is skewed. Therefore, robust nonparametric methods could be considered for this data.

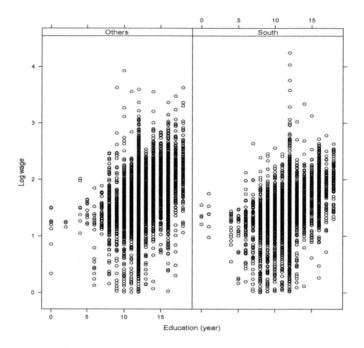

Figure 2.8 *Scatter plot of log wage against education for south and other places.*

2.2   Organization of the Book

The remainder of this book is organized as follows. Chapter 3 discusses the generalized linear model (GLM) and Quasi-likelihood method and also introduces how to model the variance the correlation. Chapter 4 mainly discusses the methods of estimating parameters for longitudinal data. Criteria for variable selection and correlation structure selection are introduced in Chapter 5. Chapter 6 discusses several roust methods for parameter estimation. Chapter 7 mainly introduces the clustered data analysis and statistical inference for intra-cluster correlation. Chapter 8 mainly focuses on missing data, including the patterns and mechanism of missing data, and how to analyze the longitudinal data with missing values. Chapter 9 introduces another two important models in longitudinal data: random-effects models and transitional models. The last chapter, Chapter 10, provides the "penalty" approach for high dimension data. This lasso-type approach has been found very effective in eliminating the redundant genes and keeping useful predictors.

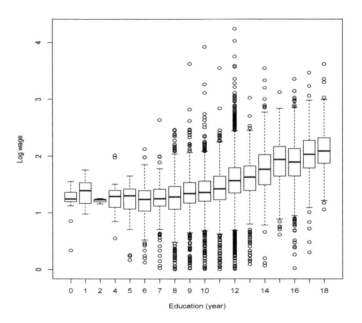

Figure 2.9 *The boxplots of the income of the women with different education time.*

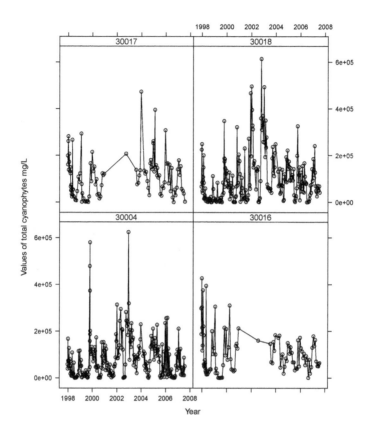

Figure 2.10 *Time series plot of the values of Total Cyanophyte from 4 sites.*

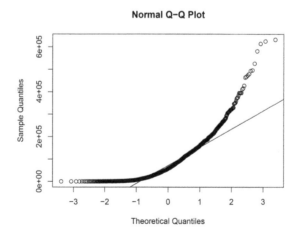

Figure 2.11 *Q-Q plot for total Cyanophyte counts from 20 sites.*

Chapter 3

Model Framework and Its Components

Suppose that data are collected from N subjects, and Y_i is the response data from subject i, where $i = 1,\ldots,N$. We first assume Y_i is univariate in this chapter, and then allow Y_i to be a vector of responses in longitudinal or clustered data analysis in Chapter 4. The predictors associated with Y_i are collected in a vector of dimension p, X_i. The design matrix for all the observations is then a matrix of dimension $p \times N$, $X = (X_1^\top, \ldots, X_N^\top)$. The response vector Y is $(Y_1, \ldots, Y_N)^\top$. Note that the Y_i is also referred to as dependent variable, output variable, or endogenous variable in econometrics, and the X_i vector is also referred to as independent variables, input variables, explanatory variables, or exogenous variables.

3.1 Distributional Theory

We are familiar with the distribution types, such as normal, Poisson, and binomial families. They are the fundamental elements in distributional theory. They are "static" in the sense that for a given parameter value, the distribution is then fixed, describing the randomness of observation at a given time. In regression or conditional modeling, we describe how Y changes accordingly when X varies. If $f(y|x_1)$ is the probability density function (pdf) of the response given $X = x_1$, how should we describe the distribution when $X = x_2$? If we believe a particular family of distributions is rich enough, $f(y|x_1)$ and $f(y|x_2)$ will correspond to two members in the distribution family associated with or determined by two parameter values. If we think of a series of observations from a number of different subjects, we may impose a set of parameter values, one for each subject. Each distribution family can be regarded as a convenient mathematical description of randomness for a particular type of data. Nevertheless, theoretically, it is possible that distributions from different subjects may **not** come from the same family. This motivates the development of the exponential distribution family, where the generalized linear models (GLMs) hinge on. A different distribution can be obtained by simply using a different exponential parameter that is determined by the linear combination of the predictors. Therefore, we can think of each family consisting of many different members, and they can be used to differentiate the distribution functions (including the mean and variance) as the predictors change.

Let Y be a random variable representing any of the responses, which is characterized by a distribution function $f(y)$. Its randomness may be contributed by many

DOI: 10.1201/9781315153636-3

factors. When we are interested in investigating the relationships between Y and X (causal or just a predictor). We shall ask the question of how Y changes when X changes. This is reflected in the conditional distribution function $f(y|x)$. For example, Y is the bodyweight of an individual from the Australian population. Its randomness is partially caused by biological difference due to X =male or female – we, therefore, may be interested in investigating the conditional distributions of Y given X, $f(y|X = Male)$, and $f(y|X = Female)$. In this case, if the normal family is used, we can allow a swift in the mean to reflect the difference in these two distributions. However, the population with a larger mean may exhibit a larger variance as well. To this end, we can either model variance just as we model the mean, or we can link the variance as a function of the mean.

Distribution families all imply a functional relationship between the mean and the variance functions due to the probabilistic regularization. If X is a continuous variable such as height in inch, $f(y|X = x)$ is then a function of x (f may not be continuous in x). Intuitively, conditioning on X explains certain variations due to X, so the variance of the residuals should shrink. As more and more useful predictors (X_i) are used, $f(y|X_i)$ should have smaller and smaller variance leading to more accurate prediction. Of course, this accurate prediction relies on our accurate description of $f(y|X_i)$, which is not easy especially when the dimension of X_i becomes large.

Exercise 1.   If var(Y) exists, prove $\mathbf{E}(\text{var}(Y|X)) \le \text{var}(Y)$ by establishing the following identity

$$\text{var}(Y) = \mathbf{E}(\text{var}(Y|X)) + \text{var}(\mathbf{E}(Y|X)).$$

Exercise 2.   Generate 1,000 normal responses from a linear model in R, y_i with mean $\mu_i = a + bx_i$ and variance $\sigma_i^2 = (\mu_i/75)^2$, where x_i is either 0 or 1 with 50% probability. Obtain the density plot for the two groups, $a = 165$ and $b = 5$. This may represent Male and Female heights (cm) from 1,000 people. How about if we wish to have a skewed distribution, what distribution can we do? Revise the R code accordingly.

In classical regression with regressors X_i, we assume $f(y|X_i)$ is normal, $N(X_i^\top \beta, \sigma^2)$. As X_i changes, the response mean is assumed to change in a linear fashion. Here X_i can be random or fixed, and the marginal density $f(y)$ is not of interest.

3.1.1   Linear Exponential Distribution Family

Suppose Y is a random variable and its distribution takes the form

$$f(y|\theta) = \exp\left\{ \frac{T(y)\theta - b(\theta)}{\phi} + c(y,\phi) \right\} dv(y), \tag{3.1}$$

where $T(y)$ is a known function such as y or y^k with a known k, θ is the natural (also known as canonical) parameter, and ϕ is a dispersion parameter, $c(y,\phi)$ is a known function of y and ϕ, and $v(y)$ is a measure function.

Example 1. For a given power parameter k, the Weibull distribution belongs to the exponential family with $T(y) = y^k$,

$$f(y|\theta) = \exp\left\{ \frac{y^k\theta - b(\theta)}{\phi} + c(y,\phi) \right\}, \quad y \in (0,\infty),$$

Table 3.1 *Components of commonly used exponential families: ϕ is the dispersion parameter, $b(\theta)$ is the cumulate function, $\mu(\theta)$ is the mean function as the natural parameter, $V(\mu)$ is the mean-variance relationship, and final row is the inverse function of $\mu(\theta)$ as the canonical link to $\eta = X^\top\beta$. If the link function $h(\mu) = X^\top\beta$ nominated is not the canonical one, $\eta = X^\top\beta$ is not the natural parameter θ any more.*

	Normal	Poisson	Binomial	Gamma	IGaussian
	$N(\mu,\sigma^2)$	$\text{Poi}(\lambda)$	$\text{Binom}(m,\pi)$	$\Gamma(\mu,\nu)$	$\text{IG}(\mu,\lambda)$
ϕ	σ^2	1	$1/m$	$1/\nu$	$1/\lambda$
$b(\theta)$	$\theta^2/2$	e^θ	$m\log(1+e^\theta)$	$-\log(-\theta)$	$-\sqrt{-2\theta}$
$\mu(\theta)$	θ	e^θ	$me^\theta/(1+e^\theta)$	$-1/\theta$	$1/\sqrt{-2\theta}$
$V(\mu)$	1	μ	$\mu(m-\mu)$	μ^2	μ^3
θ	μ	$\log(\mu)$	$\log\{\mu/(m-\mu)\}$	$1/\mu$	$-1/(2\mu^2)$

where $\theta = \lambda^{-k}$ and $b(\theta) = \log(\theta)$.

The most commonly used exponential family is the linear exponential family with $T(y) = y$. Consider a set of independent random variables $Y_1,\ldots,Y_N$, each with a distribution from the exponential family,

$$f(y;\theta,\phi) = \exp\left\{\frac{y\theta - b(\theta)}{\phi} + c(y,\phi)\right\}d\nu(y), \qquad (3.2)$$

in which θ is the natural (also known as canonical) parameter and ϕ is a dispersion parameter, and $\nu(y)$ is a measure function. Here y can take continuous or categorical data ($\nu(y)$ is a Lebesgue measure for continuous y and count measure for discrete y). Clearly, $\mathbf{E}(Y) = \mu = b'(\theta)$ and $\text{var}(Y) = \phi b''(\theta)$, both are functions of θ. It is also common to write the variance as

$$\text{var}(Y) = \phi V(\mu),$$

where $V(\cdot)$ is the so-called variance function.

In fact, $b(\theta)$ is the most important function: it determines the type of distribution (Gaussian, Poisson etc.), and its derivatives $b'(\theta)$ is actually the mean, while its second derivative determines the variance function ($V = b''(\theta)$) of the distribution. Table 3.1 lists the commonly used exponential families.

Example 2. Suppose Y follows a normal distribution $N(\mu,\sigma^2)$, where σ^2 does not change with μ,

$$f(y|\mu,\sigma^2) = \frac{1}{\sqrt{2\pi}\sigma}e^{-\frac{1}{2}\frac{(y-\mu)^2}{\sigma^2}} \quad \propto \quad e^{\frac{y\mu-\mu^2/2}{\sigma^2}}.$$

Here the proportion symbol $\propto$ is used because the constant terms free from the parameter of interest (i.e., θ in this case) are ignored. Clearly, we have $b(\theta) = \theta^2/2$, and σ^2 is the dispersion parameter.

Example 3. The inverse Gaussian distribution describes the time to the first passage of a Brownian motion and has (pdf)

$$f(y|\mu, \lambda) = \sqrt{\frac{\lambda}{2\pi y^3}} \exp\left\{\frac{-\lambda(y-\mu)^2}{2\mu^2 y}\right\}$$

$$\propto \exp\left\{\lambda(y\theta - \sqrt{-2\theta})\right\},$$

in which the natural parameter $\theta = -1/(2\mu^2)$ is always negative and $b(\theta) = -\sqrt{-2\theta}$. Therefore, under the canonical link, the linear score $\eta = X^\top \beta$ must be negative.

Example 4. Let us now consider a general power function including two limiting cases for Gamma and Poisson distributions,

$$b(\theta) = \begin{cases} \left(\frac{\theta}{\tau-1}\right)^\tau \frac{\tau-1}{\tau} & \text{if } \tau \neq 0 \text{ or } 1 \\ -\log(-\theta) & \text{if } \tau = 0 \text{ (Gamma distribution).} \\ e^\theta & \text{if } \tau \to \infty \text{ (Poisson distribution)} \end{cases}$$

The normal and inverse Gaussian distributions are included when $\tau = 2$ and $\tau = -0.5$. The distributions corresponding to other τ values are known as Tweedie exponential family (Tweedie, 1984; Jørgensen, 1997). The power relationship between the mean $\mu = b'(\theta)$ and variance $b''(\theta)$ is seen when $\{\theta/(\tau-1)\}^\tau(\tau-1)/\tau$, where $\tau \neq 1$ and can be positive or negative. Note the limiting value of $b(\theta)$ is $\log(-\theta)$ when $\tau = 0$. It is interesting to see $\tau = 2$, 0, and -0.5 correspond to normal, Gamma, and inverse Gaussian distributions.

3.1.2   Quadratic Exponential Distribution Family

A straightforward generalization to the linear exponential family is to include y^2 term in (3.2),

$$f(y|\theta) = \exp\left\{\frac{y\theta + y^2\lambda - b(\theta, \lambda)}{\phi} + c(y, \phi)\right\} dv(y), \qquad (3.3)$$

in which both θ and λ will be functions of the mean and variance, and

$$\frac{\partial b(\theta, \lambda)}{\partial \theta} = \mu, \quad \frac{\partial b(\theta, \lambda)}{\partial \lambda} = E(Y^2),$$

and

$$\frac{\partial^2 b(\theta, \lambda)}{\partial \theta^2} = \sigma^2, \quad \frac{\partial^2 b(\theta, \lambda)}{\partial \lambda^2} = \text{var}(Y^2).$$

More details can be found in the book by Ziegler (2011, Ch. 2).

3.1.3   Tilted Exponential Family

Consider a regression approach when we are interested in investigating the effect of X_i — how Y_i changes as X_i changes. Suppose we use X_0 as the reference values (such as the Male group, or average age etc.) and the corresponding response variable Y_i

has the distribution, $f(y|X_0^\top\beta)$, which is the reference distribution. To model the effect of X_i, we only need to look at the ratio

$$f(y|X_i^\top\beta)/f(y|X_0^\top\beta).$$

Under the linear exponential family framework, this ratio is

$$f(y|X_i^\top\beta)/f(y|X_0^\top\beta) = \exp\left\{\frac{(\theta_i - \theta_0)y - b(\theta_i) + b(\theta_0)}{\phi} + c(y,\phi)\right\}. \tag{3.4}$$

This leads to the tilted distribution family (Rathouz and Gao, 2009),

$$f(y|X_i^\top\beta) = \exp(b_i + \theta_i y)f(y|X_0^\top\beta),$$

where b_i is the normalizing constant so that $\int f(y|X_i^\top\beta)dy = 1$, and θ_i is so chosen that the mean assumption is met, $\int yf(y|X_i^\top\beta)dy = \mu_i$. This generalizes the linear exponential family in the sense that $f(y|X_0^\top\beta)$ is unspecified and does not have to take the linear exponential distribution. Huang and Rathouz (2012) proposed semiparametric and empirical likelihood approach for estimating β.

This tilted model bears some similarity to the proportional hazard model (Cox, 1972), where the hazard ratio $f(y)/(1 - F(y))$ instead of $f(y)$ is tilted. The exponential distribution family provides a family of distributions indexed by the canonical parameter values, but the family does not specify how these family members are linked. The GLM hinges on the exponential family with a link function specifying how the canonical values are linked to the covariates. The variance is still the derivation of the mean with respect to θ, but no explicit forms exist in general for either the mean or the variance functions.

The tilted exponential family provides more than just a family of distributions, and the "canonical" parameter value is implicitly defined by the desired marginal mean via the link function. The "canonical" parameter here plays a tilting role so that the reference distribution is tilted as the distribution at the new mean values. Much more work is needed on this topic for analysis of longitudinal data.

3.2 Quasi-Likelihood

When the true likelihood is difficult to specify, we can then rely on the quasi-likelihood (QL) function proposed by Wedderburn (1974). Instead of claiming the true distribution of y_i is known, we only need to model the first two moments of the distribution and construct a "likelihood" alike function (the so-called quasi-likelihood) to work with. In the regression context, we specify only the relationship between the mean and covariates and the variance function as the mean (for weighting).

We have assumed that Y is a random variable with mean $\mu = h(X^\top\beta)$ and variance $\phi V(\mu)$. The QL (on log scale) for Y_i is defined as (Wedderburn, 1974)

$$Q(\mu_i; y_i) = \int_{y_i}^{\mu_i} \frac{Y_i - t}{\phi V(t)} dt.$$

The overall QL for all the independent observations, $Y = (Y_1, \ldots, Y_N)$, is defined as

$$Q(\mu; Y) = \sum_{i=1}^{N} Q(\mu_i; Y_i),$$

where μ is the mean vector.

Remark 1:   The QL is a likelihood function, but it is not meant to be the true likelihood that generates the observed data.

Remark 2:   The QL function is more than just a "working" likelihood in the sense that it is meant to be an approximation to the true likelihood when only the first two moments are matched.

Remark 3:   The QL function can be treated as the true likelihood is the sense that the resultant score functions and the information matrix are valid.

Remark 4:   The true score function may not be a linear combination of the data (for example, y_i^2 can be involved), and in that case QL is still valid for estimating β and deriving the information matrix. Of course, if the true likelihood is used, the MLE will be more efficient but at the cost that potential biases exist in $\hat{\beta}$ when the likelihood is misspecified although the first two moments are correctly matched.

Remark 5:   In the case the variance function $V(\mu)$ is incorrect, the QL score function is valid for estimating β, but the information matrix becomes invalid.

For dependent observations, multivariate analysis will be used. The QL can be defined in a similar way,

$$Q(\mu; Y) = \int \int_{t(s)} (Y - t)^{\top} V^{-1}(t) dt(s),$$

where $t(s)$ is a smooth curve of dimension n joining Y and μ. For this integral to be meaningful as a likelihood, it needs to be path-independent (§9.3, McCullagh and Nelder, 1989). For longitudinal data analysis, the joint likelihood for the n random variables from some subject can be written in the form of n products as,

$$f(y_1, y_2, \ldots, y_n) = f(y_1)f(y_2|y_1)f(y_3|y_1, y_2) \ldots f(y_n|y_1, y_2, \ldots, y_{n-1}). \tag{3.5}$$

Wang (1996) constructed the quasi-likelihood (QL) for $f(y_j|y_1, y_2, \ldots, y_{j-1})$ in the context of categorical data analysis. An interesting application was done by Wang (1999a) for estimating a population size from removal data. The conditional quasi likelihood was constructed via the conditional mean and conditional variance of the catch given previous total catches.

Remark 6:   This approach can be easily extended for any multivariate variables by modeling the conditional mean and variance. The resultant QL function is, in general, not a scaled version of the ordinary log-likelihood function.

3.3   Gaussian Likelihood

Let us first consider the independence case assuming $n_i = 1$. In the lack of the likelihood function that generates the data, one may pretend that y_i were generated from a normal distribution $N(\mu_i, \sigma_i^2)$, in which $\sigma_i^2 = \sigma^2 g(\mu_i)$, where $\mu_i = h(X_i\beta)$. This is known as the pseudolikelihood approach. The -2 log-likelihood function is then

$$L_{G0}(\beta; \gamma) = \sum_{i=1}^{N} \left(\frac{y_i - \mu_i}{\sigma_i} \right)^2 + \sum_{i=1}^{N} \log(\sigma_i^2). \tag{3.6}$$

The question is to which extent the estimation and inference are still valid. It is amazing how valid this approach can be even for binary data. Whittle (1961) and Crowder (1985) introduced the Gaussian estimation as a vehicle for estimation without requiring that the data are normally distributed. The function above may also be called Pseudo-likelihood (Davidian and Carroll, 1987). For longitudinal data analysis, the Gaussian approach can also be applied (Crowder, 1985, 2001; Wang and Zhao, 2007). Recall that the scalar response y_{ij} is observed for cluster i ($i = 1, \ldots, N$), at time j ($j = 1, \ldots, n_i$). For the ith cluster, let $Y_i = (y_{i1}, \ldots, y_{in_i})^T$ be a $n_i \times 1$ response vector, and $\mu_i = \mathbf{E}(Y_i)$ is also a $n_i \times 1$ vector. We denote $\text{Cov}(Y_i)$ by Σ_i, which has the general form ϕV_i, where $V_i = A_i^{1/2} R_i A_i^{1/2}$, with $A_i = \text{diag}\{\text{Var}(y_{it})\}$ and R_i being the correlation matrix of Y_i. For independent data, Σ_i is just ϕA_i.

The Gaussian log-likelihood (multiplied by -2) for the data $(Y_1, \ldots, Y_N)$, is

$$L_G(\beta; \tau) = \sum_{i=1}^{N} \log\{|\Sigma_i| + (Y_i - \mu_i)^\top \Sigma_i^{-1}(Y_i - \mu_i)\}, \tag{3.7}$$

where β is the vector of regression coefficients governing the mean and τ is the vector of additional parameters needed to model the covariance structure realistically. Later on we will let θ collect all the parameters including both β and τ. Then, we can write μ_i and Σ_i in a parametric form, respectively: $\mu_i = \mu_i(\beta)$ and $\Sigma_i = \Sigma_i(\beta, \tau)$. The Gaussian estimation is performed by maximizing $L_G(\theta)$ over θ.

Exercise 3. Suppose independent data $(y_1, \ldots, y_N)$ are generated from the distribution function (3.1). Prove that $\sum_{i=1}^{N} T(y_i)$ is a sufficient statistic for θ.

Exercise 4. For the exponential distribution family given by (3.2), we can show that $\mathbf{E}(Y) = b'(\theta)$, and $\text{var}(Y) = \phi b''(\theta)$.

Exercise 5. Suppose Y is a compound distribution in the sense that $Y|\theta \sim$ Poisson(θ) and $\theta|\lambda$ is also Poisson(λ). Show that

(a). $\text{Var}(Y) = 2\lambda$.
(b). Work out the probability $P(Y = 5)$.
(c). Verify $\sum_{i=0}^{\infty} P(Y = i) = 1$.
(d). Verify $\sum_{i=0}^{\infty} i P(Y = i) = \lambda$.
(e). What is $\sum_{i=0}^{\infty} i^2 P(Y = i)$?

3.4 GLM and Mean Functions

The exponential distribution family is quite general and flexible in describing a distribution. This is great news for modeling independent and identically distributed (i.i.d.) data which rarely comes by in practice. For example, if we are interested in investigating the effect of X_i on the response variable Y_i, it would be reasonable to impose i.i.d. assumption on the responses from subjects who have the same X_i. However, it is critical to describe how the distribution changes when X_i values differ.

To this end, a regression framework can be adopted. The GLM assumes a relationship between the mean μ_i and X_i via a link function, $g(\mu_i) = X_i^\top \beta$. For a given X_i values, the distribution of y_i is then determined based on (3.2) by adjusting the natural parameter θ to match the desired mean via the inverse function of g such that $\mu_i = g^{-1}(X_i\beta)$.

The following diagram shows the skeleton of the GLM: the first part from θ to $b'(\theta)$ specifies the distribution of each response y_i, and the second mapping is the link function describing how the mean changes as the covariate X_i changes.

$$\theta_i \overset{b'(\theta_i)}{\to} \mu_i \overset{g(\mu_i)}{\to} X_i\beta.$$

If we wish to use the linear score $X_i\beta$ be the natural parameter θ, the compounding effect of $b'(\theta)$ and $g(\mu_i)$ must become the identity function (i.e. $g(b'(\theta)) = \theta$) so that $X_i\beta = \theta$. In this case, the link $g(\cdot)$ is, therefore, automatically determined when $b(\theta)$ is known.

Now let us consider the overall log-likelihood for (Y,X),

$$L = \phi^{-1} \sum_{i=1}^{N} \{Y_i\theta_i - b(\theta_i)\} + \sum_{i=1}^{N} c(Y_i,\phi)d\mu_i.$$

Denote the linear score $X_i\beta$ as η_i and $G(\eta) = \partial\theta/\partial\eta_i$, which is 1 when canonical link is used. The corresponding score function is

$$\sum_{i=1}^{N} X_i G(\eta_i)(Y_i - \mu_i). \tag{3.8}$$

As we can see, under canonical link, the estimating functions take the following simple form,

$$S(\beta) = \sum_{i=1}^{N} X_i(Y_i - \mu_i), \tag{3.9}$$

which is a linear combination of the residuals $Y_i - \mu_i$.

When the link function is not canonical, the log-likelihood function may not be convex in β and finding the MLE may incur numerical problems. This can also be seen in (3.8) as the unknown parameters also appear in $G(\cdot)$, which can make the estimating functions nonmonotonic even if μ_i is monotonic in β.

For the likelihood-based estimates, the covariance of $\hat{\beta}$ is given by the inverse of fisher information matrix, regardless the canonical link is used or not,

$$-\mathbf{E}(\partial^2 L/\partial\beta\partial\beta^\top).$$

However, if (3.9) is used as estimating functions for solving β and the canonical link is not used because either a noncanonical function is used for the mean function or we are not sure if the data come from the linear exponential family, we will have

$$\hat{\beta} - \tilde{\beta} \sim N\left(0, \left(\sum_{i=1}^{N} X_i D_i^\top\right)^{-1} \left[\sum_{i=1}^{N} X_i \mathrm{var}(Y_i)X_i^\top\right]^{-1} \left(\sum_{i=1}^{N} D_i X_i^\top\right)^{-1}\right).$$

Note that var(Y_i) is usually unknown and will be estimated either by a variance model or $(Y_i - \hat{\mu}_i)^2$. In the latter, the sample size N must be large enough so that the limiting theorem can be applied to approximate $\sum_{i=1}^{N} D_i^\top \epsilon_i^2 D_i / N \approx \sum_{i=1}^{N} \mathbf{E}(D_i^\top \epsilon_i^2 D_i)/N$.

Example 5. Poisson regression with a canonical or noncanonical link

Suppose $Y_i|X_i \sim \text{Poi}(\mu_i)$, and the log-likelihood function is

$$\sum_{i=1}^{N} \{Y_i \log(\mu_i) - \mu_i\}.$$

If the canonical link is used, $\log(\mu_i) = X_i\beta$, we have the following score functions for β,

$$\sum_{i=1}^{N} X_i(Y_i - e^{X_i\beta}).$$

Now let us consider a noncanonical link, $\theta = \log(\mu_i) = \exp(X_i\beta)$, for example. The resultant score functions take the form

$$\sum_{i=1}^{N} X_i e^{X_i\beta} \{Y_i - e^{X}(e^{X_i\beta})\}.$$

Example 6. Suppose Y follow a gamma distribution $\Gamma(\alpha,\beta)$,

$$f(y|\alpha,\beta) \propto y^{\alpha-1} e^{-y/\beta} = e^{X}\{-y/\beta + (\alpha-1)\log(y)\}.$$

Clearly, for a given α, $f(y|\alpha,\beta)$ belongs to the linear exponential family with the natural parameter $\theta = -\beta^{-1}$. On the other hand, if β is known (or does not change with mean and treated as a nuisance parameter), $f(y|\alpha,\beta)$ belongs to the nonlinear exponential distribution family with $\theta = (\alpha - 1)$ and $T(y) = \log(y)$. For this gamma distribution, we know $\mu_i = \alpha\beta$ and $\sigma_i^2 = \alpha\beta^2 = \mu_i\beta = \mu_i^2/\alpha$. This indicates that the variance can be modeled as proportional to μ_i when fixing β (nonlinear exponential distribution with $T(y) = \log(y)$) or as proportional to μ_i^2 when fixing α (linear exponential distribution).

How about other power relationships? Suppose we desire to have $\sigma_i^2 \propto \mu_i^\gamma$ to model variance heterogeneity. This can be achieved when α changes with β in a way $\alpha = \beta^{(\gamma-2)/(\gamma-1)}$.

Exercise 6. Prove that (3.8) coincides with the result when applying the Gauss-Markov theorem to the residuals $Y_i - \mu_i$.

Exercise 7. Suppose that y_i is generated from the normal distribution $N(\mu_i, \sigma^2\mu_i^2)$, where $\mu_i = X_i\beta$. Obtain the score functions for β. Explain why they do not have the form as (3.9) or (3.8). Are there any link functions $\mu_i = g(X_i\beta)$ so that the resultant score functions take the form of (3.9) or (3.8)?

Unlike the classical linear regression model, which can only handle the normally distributed data, GLM extends the approach to count data, binary data, continuous data which need not be normal. Therefore, GLM is applicable to a wider range of data analysis problems.

Under the assumption that $\mathbf{E}(Y_i) = \mu_i$ regardless Y_i is continuous or not, we can consider the following regression model for estimating the parameters β in the mean function μ_i,

$$Y_i = \mu_i + \epsilon_i.$$

Here ϵ_i is simply the difference between the observed Y_i and its expectation μ_i. The ordinary least squares approach leads to the following estimating function in which D_i is the Jacobian matrix $\partial \mu_i / \partial \beta$,

$$\sum_{i=1}^{N} D_i^\top \{Y_i - \mu_i(\beta)\}. \tag{3.10}$$

To account for possible variance heterogeneity in Y_i, a weighted least squares (WLS) can be adopted, but a known weight (w_i, say) must be supplied, and the resultant estimating function is

$$\sum_{i=1}^{N} D_i^\top w_i \{Y_i - \mu_i(\beta)\}. \tag{3.11}$$

In general, w_i is the reciprocal of the variance of Y_i (up to a proportion constant) and unknown. In the GLM setting, we have $\sigma_i^2 = h(\mu_i)$ and may wish to use $w_i = 1/h(\hat{\mu}_i)$ for some estimated $\hat{\mu}_i$ in the WLS approach. This is the key idea in the iterative weighted least squares (IWLS), which iterates between estimating $\hat{\beta}$ and updating $\sigma_i^2 = h(\hat{\mu}_i)$ as w_i. At convergence, the $\hat{\beta}$ is the same as those from the following estimating function,

$$\sum_{i=1}^{N} D_i^\top \sigma_i^{-2}(\beta) \{Y_i - \mu_i(\beta)\}. \tag{3.12}$$

Exercise 8. Suppose $\hat{\beta}_w$ is obtained from (3.12) and $\hat{\beta}_{\text{ora}}$ is an oracle estimator obtained by solving

$$\sum_{i=1}^{N} D_i^\top \sigma_i^{-2}(\tilde{\beta}) \{Y_i - \mu_i((\beta)\},$$

in which the weighting is based on the true $\tilde{\beta}$.
Prove (a). $\hat{\beta}_w - \tilde{\beta} = O_p(N^{-1/2})$.
(b). $\hat{\beta}_{\text{ora}} - \tilde{\beta} = O_p(N^{-1/2})$.
(c). $\hat{\beta}_w - \hat{\beta}_{\text{ora}} = o_p(N^{-1/2})$.

Exercise 9. Show that (3.12) is the score function if Y_i is from the linear exponential distribution family.

3.5 Marginal Models

The marginal models aim to make inferences at the population level. This approach is to model the marginal distributions $f(y_{ij})$ at each time point, for $j = 1, 2, \ldots, n_i$, instead of the joint distributions for all the repeated measures.

The emphasis is often on the mean (and variance) of the univariate response y_{ij} the population level instead of on the individual i at any given fixed covariates. A covariance structure is often nominated to account for the within-subject correlations.

A feature of marginal models is that the models for the mean and the covariance are specified separately. Marginal models are considered a natural approach when we wish to extend the generalized linear model methods for the analysis of independent observations to the setting of correlated responses.

Specification of a mean function is the premier task in the GEE regression model. If the mean function is not correctly specified, the analysis will have no meaning.

Under the work frames of GLM, the link function provides a link between the mean and a linear combination of the covariates. The link function is called canonical link if the link function equals to the canonical parameters. Different distribution models are associated with different canonical links. For Normal, Poisson, Binomial, and Gamma random components, the canonical links are identity, log-link, logit-link, and inverse link, respectively.

We assume that the response variable, y_{ki}, has mean μ_{ij} and variance $\phi \sigma_{ij}^2$. Here ϕ is an unknown scale parameter, μ_{ij} and σ_{ij}^2 are some known functions of the covariates, and X_{ij} is a $p \times 1$ vector. Let $\mu_i = (\mu_{ij})$ be the marginal mean vector for subject i. The covariance matrix of Y_i, Σ_i, is assumed to have the form $\phi A_i^{1/2} R_i(\alpha) A_i^{1/2}$, in which $A_i = \text{diag}(\sigma_{ij}^2)$ and $R_i(\alpha)$ is the correlation matrix parameterized by α, a q-vector.

We first consider the special case when $n_i = 1$ and all observations are independent of each other. In generalized linear models (GLMs), we have $\mu_{ij} = h(X_{ij}^\top \beta)$ where $h(\cdot)$ is the inverse function of the link function., a function of some linear combination of the covariates, and the variance σ_{ij}^2 is related to the covariates via some *known* function of the mean.

The counterpart of the random effect model is a marginal model. A marginal model is often used when inference about population averages is of interest. The mean response modeled in a marginal model is conditional only on covariates and not on random effects. In marginal models, the mean response and the covariance structure are modeled separately.

We assume that the marginal density of y_{ij} is given by,

$$f(y_{ij}) = e^X[\{y_{ij}\theta_{ij} - b(\theta_{ij})\}/\phi + c(y_{ij}, \phi)].$$

That is, each y_{ij} is assumed to have a distribution from the exponential family. Specifically, with marginal models we make the following assumption:

• The marginal expectation of the response, $\mathbf{E}(y_{ij}) = \mu_{ij}$, depends on explanatory variables, X_{ij}, through a known link function, the inverse function of h,

$$h^{-1}(\mu_{ij}) = \eta_{ij} = X_{ij}^\top \beta.$$

• The marginal variance of y_{ij} is assumed to be a function of the marginal mean,

$$\text{var}(y_{ij}) = \phi v(\mu_{ij}),$$

in which $v(\mu_{ij})$ is a known "variance function", and ϕ is a scale parameter that may need to be estimated.

• The correlation between y_{ij} and y_{ik} is a function of some covariates (usually just time) with a set of additional parameters, say α, that may also need to be estimated.

Here are some examples of marginal models:

• Continuous responses:
1. $\mu_{ij} = \eta_{ij} = X_{ij}^\top \beta$ (i.e. linear regression), identity link
2. $\text{var}(y_{ij}) = \phi$ (i.e. homogeneous variance)
3. $\text{Corr}(y_{ij}, y_{ik}) = \alpha^{|k-j|}$ (i.e. autoregressive correlation)

• Binary response:
1. $\text{Logit}(\mu_{ij}) = \eta_{ij} = X_{ij}^\top \beta$ (i.e. logistic regression), logit link
2. $\text{var}(y_{ij}) = \mu_{ij}(1 - \mu_{ij})$ (i.e. Bernoulli variance)
3. $\text{Corr}(y_{ij}, y_{ik}) = \alpha_{jk}$ (i.e. unstructured correlation)

• Count data:
1. $\log(\mu_{ij}) = \eta_{ij} = x_{ij}^\top \beta$ (i.e. Poisson regression), log link
2. $\text{var}(y_{ij}) = \phi \mu_{ij}$ (i.e. extra-poisson variance)
3. $\text{Corr}(y_{ij}, y_{ik}) = \alpha$ (i.e. compound symmetry correlation)

3.6   Modeling the Variance

When analyzing count data, we often assume the variance structure is the one of Poisson distribution, that is $\text{Var}(y) = \mathbf{E}(y) = \mu$. But for some count data, such as epileptic seizures data mentioned previously, the variance structure $\text{Var}(y) = \mu$ seems inappropriate, because the sample variance is much larger than the sample mean. Misspecification of variance structure will lead to the low efficiency of regression parameter estimation in longitudinal data analysis. One sensible way is to use a different variance function according to the features of the data set. Many variance functions, such as exponential, extra Poisson, powers of μ, have been proposed in Davidian and Giltinan (1995).

Here we consider the variance function as a power function of μ:

$$\mathbf{V}(\mu) = \mu^\gamma.$$

Most common values of γ are the values of 0, 1, 2, 3, which are associated with Normal, Poisson, Gamma, and Inverse Gaussian distributions, respectively. K. (1981) also discussed distributions with this power variance function and showed that an exponential family exists for $\gamma = 0$ and $\gamma \geq 1$. In Jørgensen (1997), the author summarized Tweedie exponential dispersion models and concluded that distributions do not exist for $0 < \gamma < 1$. For $1 < \gamma < 2$, it is compound Poisson; For $2 < \gamma < 3$ and $\gamma > 3$, it is positive stable distribution. The Tweedie exponential dispersion model is denoted $Y \sim Tw_\gamma(\mu, \phi)$. By definition, this model has mean μ and variance

$$\text{Var}(Y) = \phi \mu^\gamma.$$

Now we try to find the exponential dispersion model corresponding to $V(\mu) = \mu^\gamma$. The exponential dispersion model extends the natural exponential families and includes many standard families of distribution.

However, from the likelihood perspective, it is interesting to find what other likelihood functions can lead to such variance functions. Denote exponential dispersion model with $ED(\mu, \phi)$, and it has the following distribution form:

$$\exp[\lambda\{y\theta - \kappa(\theta)\}]\upsilon_\lambda(dy),$$

where υ is a given σ-finite measure on $\mathcal{R}$ (the set of all real numbers). The parameter θ is called the canonical parameter, λ is called the index parameter (and $\phi = 1/\lambda$ is called the dispersion parameter). The parameter μ is called the mean value parameter.

The cumulative generating function of $Y \sim ED(\mu, \phi)$ is

$$K(s; \theta, \lambda) = \lambda\kappa(\theta + s/\lambda) - \kappa(\theta).$$

Let κ_γ and τ_γ denote the corresponding unit cumulative function and mean value mapping, respectively. For exponential dispersion models, we have the following relations

$$\frac{\partial \tau_\gamma^{-1}}{\partial \mu} = \frac{1}{V_\gamma(\mu)}$$

and

$$\kappa_\gamma'(\theta) = \tau_\gamma(\theta).$$

If the exponential dispersion model corresponding to V_γ exists, we must solve the following two differential equations,

$$\frac{\partial \tau_\gamma^{-1}}{\partial \mu} = \mu^{-\gamma}, \tag{3.13}$$

and

$$\kappa_\gamma'(\theta) = \tau_\gamma(\theta). \tag{3.14}$$

It is convenient to introduce the parameter φ, defined by

$$\varphi = \frac{\gamma - 2}{\gamma - 1}, \tag{3.15}$$

with inverse relation

$$\gamma = \frac{\varphi - 2}{\varphi - 1}. \tag{3.16}$$

From (3.13) we find,

$$\tau_\gamma(\theta) = \begin{cases} \left(\frac{\theta}{\varphi - 1}\right)^{\varphi - 1} & \text{if } \gamma \neq 1 \\ e^\theta & \text{if } \gamma = 1 \end{cases}.$$

From τ_γ we find κ_γ by solving (3.14), which gives,

$$\kappa_\gamma(\theta) = \begin{cases} \frac{\varphi-1}{\varphi}(\frac{\theta}{\varphi-1})^\varphi & \text{if } \gamma \neq 1,2 \\ e^\theta & \text{if } \gamma = 1 \\ -\log(-\theta) & \text{if } \gamma = 2 \end{cases}.$$

In both (3.13) and (3.14), we have ignored the constants in the solutions, which does not affect the results.

If an exponential dispersion model corresponding to (3.6) exists, the commutative generating function of the corresponding convolution model is,

$$K_\gamma(s;\theta,\lambda) = \begin{cases} \lambda\kappa_\gamma(\theta)\{(1+\frac{s}{\theta\lambda})^\varphi - 1\} & \text{if } \gamma \neq 1,2 \\ \lambda e^\theta\{\exp(s/\lambda) - 1\} & \text{if } \gamma = 1 \\ -\lambda\log(1+\frac{s}{\theta\lambda}) & \text{if } \gamma = 2 \end{cases}.$$

We now consider the case $\alpha < 0$, corresponding to $1 < \gamma < 2$. It shows that the Tweedie model with $1 < \gamma < 2$ is a compound Poisson distributions.

Let N and $X_1,\ldots,X_N$ denote a sequence of independent random variables, such that N is Poisson distributed Poisson(m) and the X_is are identically distributed. Define

$$Z = \sum_{i=1}^{N} X_i, \tag{3.17}$$

where Z is defined as 0 for $N = 0$. The distribution (3.17) is a compound Poisson distribution. Now we assume that $m = \lambda\kappa_\gamma(\theta)$ and $X_i \sim \Gamma(\varphi/\theta, -\varphi)$. Note that, by the convolution formula, we have $Z|N = n \sim \Gamma(\varphi/\theta, -n\varphi)$. The moment generating function of Z is

$$\mathbf{E}(e^{sZ}) = \exp[\lambda\kappa_\gamma(\theta)\{(1 + s/\theta)^\varphi - 1\}].$$

This shows that Z is a Tweedie model. We can obtain the joint density of Z and N, for $n \geq 1$ and $z > 0$,

$$\begin{aligned} p_{Z,N}(z,n;\theta,\lambda,\varphi) &= \frac{(-\theta)^{-n\varphi}m^n z^{-n\varphi-1}}{\Gamma(-n\varphi)n!}\exp\{\theta z - m\} \\ &= \frac{\lambda^n \kappa_\gamma^n(-1/z)}{\Gamma(-n\varphi)n!z}\exp\{\theta z - \lambda\kappa_\gamma(\theta)\}. \end{aligned}$$

The distribution of Z is continuous for $z > 0$, and summing over n, the density of Z is

$$p(z;\theta,\lambda,\varphi) = \frac{1}{z}\sum_{n=1}^{\infty}\frac{\lambda^n \kappa_\gamma^n(-1/z)}{\Gamma(-n\varphi)n!}\exp\{\theta z - \lambda\kappa_\gamma(\theta)\}. \tag{3.18}$$

Let $y = z/\lambda$, then y has the probability density function given by

$$p(y;\theta,\lambda,\varphi) = c_\gamma(y;\lambda)\exp[\lambda\{\theta y - \kappa_\gamma(\theta)\}], \; y \geq 0, \tag{3.19}$$

where

$$c_\gamma(y;\lambda) = \begin{cases} \frac{1}{y}\sum_{n=1}^{\infty}\frac{\lambda^n \kappa_\gamma^n(-\frac{1}{\lambda y})}{\Gamma(-\varphi n)n!} & y > 0 \\ 1 & y = 0 \end{cases}. \tag{3.20}$$

It is not clear how valid the likelihood functions are in statistical inferences when the data generated by a different function meeting certain moment assumptions (mean and variance, for example). As one can see, it is not straightforward to develop multivariate versions of such likelihood functions in which a dependency structure or a correlation structure is imbedded somehow. These are topics of future work.

3.7 Modeling the Correlation

Arguably, correlation modeling is the key in dependent data analysis such as time series and longitudinal data analysis. The general approach to model dependence in longitudinal studies takes the form of a patterned correlation matrix $R(\alpha)$ with $q = \dim(\alpha)$ correlation parameters.

The number of correlation parameters and the estimator of α varies from case to case in the literature. Most researchers follow Liang and Zeger (1986), who discussed a number of important special cases. We first assume that each subject has an equal number of observations ($n_i = n$). The following are the typical "working" correlation structures and the estimators used to estimate the "working" correlations.

- m-**dependent correlation**

$$\text{Corr}(y_{ij}, y_{ij+t}) = \begin{cases} 1 & t = 0 \\ \alpha_t & t = 1, \ldots, m \\ 0 & t > m \end{cases} \quad \text{and}$$

$$R = \begin{pmatrix} 1 & \alpha_1 & \alpha_2 & \cdots & \alpha_m & 0 & \cdots & 0 \\ \alpha_1 & 1 & \alpha_1 & \cdots & \alpha_{m-1} & \alpha_m & \cdots & 0 \\ \vdots & \vdots & & & & & & \\ 0 & 0 & \cdots & \cdots & \alpha_2 & \alpha_1 & \cdots & 1 \end{pmatrix}.$$

The case of $m = 1$ corresponds to the moving average (MA) structure in time series modeling. Note that α_1 is the assumed or imposed correlation of $corr(y_{11}, y_{12})$, $corr(y_{12}, y_{13})$ ($i = 1$), and also $corr(y_{21}, y_{22})$, $corr(y_{22}, y_{23})$ ($i = 2$), etc. This structure may be unreasonable if the observation times t_{ij} are $1, 2, 4, \ldots$ for $i = 1$ and $2, 3, 4$ for $i = 2$.

- **Exchangeable:**

$$\text{Corr}(y_{ij}, y_{ik}) = \begin{cases} 1 & j = k \\ \alpha & j \neq k \end{cases}.$$

- **Autoregressive correlation, AR(1)**

$$corr(Y_{ij}, Y_{i,j+t}) = \alpha^t, \quad t = 0, 1, 2, \ldots, n - j.$$

Correlation matrix is

$$R = \begin{pmatrix} 1 & \alpha & \alpha^2 & \cdots & \alpha^{n-1} \\ \alpha & 1 & \alpha & \cdots & \alpha^{n-2} \\ \alpha & \alpha^2 & 1 & \cdots & \alpha^{n-3} \\ \vdots & \vdots & \vdots & & \vdots \\ \alpha^{n-1} & \alpha^{n-2} & \alpha^{n-3} & \cdots & 1 \end{pmatrix}.$$

- **Toeplitz correlation ($n-1$ parameters)**

$$\text{Corr}(y_{ij}, y_{ik}) = \begin{cases} 1 & j = k \\ \alpha_{|j-k|} & j \neq k \end{cases} \quad \text{and} \quad R = \begin{pmatrix} 1 & \alpha_1 & \alpha_2 & \cdots & \alpha_{n-1} \\ \alpha_1 & 1 & \alpha_1 & \cdots & \alpha_{n-2} \\ \alpha_2 & \alpha_1 & 1 & \cdots & \alpha_{n-3} \\ \vdots & \vdots & \vdots & & \vdots \\ \alpha_{n-1} & \alpha_{n-2} & \alpha_{n-3} & \cdots & 1 \end{pmatrix}.$$

- **Unstructured correlation**

$$\text{Corr}(y_{ij}, y_{ik}) = \begin{cases} 1 & j = k \\ \alpha_{jk} & j \neq k \end{cases}.$$

The number of correlation parameter varies according to different "working" correlation structures. The exchangeable (uniform or compound) correlation structure has

$$R = (1-\rho)I + \rho \mathbf{e}\mathbf{e}^\top,$$

where $-1/(n_i - 1) < \rho < 1$ is unknown, $\mathbf{e} = (1, 1, \ldots, 1)^T$, and I is the identity matrix of order n_i. The serial covariance structure has:

$$\Sigma = \sigma^2 C,$$

where $C = (\rho^{|i-j|})$, and $\sigma^2 > 0$ and $-1 < \rho < 1$ are unknown. Lee (1988) studied these two correlation structures in the context of growth curves.

Note that these models are most suitable when all subjects have the same, and equal observation times, for example, each subject is observed five times (Monday to Friday). However, if Subject 1 is observed on Monday, Tuesday, and Friday; and Subject 2 is observed on Monday, Wednesday only, the corresponding two correlation matrices using AR(1) structure are

$$\begin{pmatrix} 1 & \alpha & \alpha^4 \\ \alpha & 1 & \alpha^3 \\ \alpha^4 & \alpha^3 & 1 \end{pmatrix} \quad \text{and} \quad \begin{pmatrix} 1 & \alpha^2 \\ \alpha^2 & 1 \end{pmatrix}.$$

If observation times are not a lattice nature, but rather in a continuous time, exponential correlation structure is more appropriate (Diggle, 1988), $\rho(|t_j - t_i|)$, where $\rho(u) = \exp(-\alpha u^c)$, with $c = 1$ or 2. The case of $c = 1$ is the continuous-time analog of a first-order autoregressive process. The case of $c = 2$ corresponds to an intrinsically smoother process. The covariance structure can handle irregularly spaced time sequences within experimental units that could arise through randomly missing data or by design. Besides the aforementioned covariance structures, there are still parametric families of covariance structures proposed to describe the correlation of many types of repeated data. They can model quite parsimoniously a variety of forms of dependence and accommodate arbitrary numbers and spacings of observation times, which need not be the same for all subjects. Núñez Anton and Woodworth

(1994) proposed a covariance model to analyze unequally spaced data when the error variance-covariance matrix has a structure that depends on the spacing between observations. The covariance structure depends on the time intervals between measurements rather than the time order of the measurements. The main feature of the structure is that it involves a power transformation of the time rather than time interval and the power parameter is unknown.

The general form of the covariance matrix for a subject with k observation at times $0 < t_1 < t_2 < \ldots < t_k$ is

$$(\Sigma)_{uv} = (\Sigma)_{vu} = \begin{cases} \sigma^2 \cdot \alpha^{(t_v^\lambda - t_u^\lambda)}/\lambda & \text{if } \lambda \neq 0 \\ \sigma^2 \cdot \alpha^{\log(t_v/t_u)} & \text{if } \lambda = 0. \end{cases}$$

($1 \leq u \leq v \leq k, 0 < \alpha < 1$). The covariance structure consists of three-parameter vector $\theta = (\sigma^2, \alpha, \lambda)$. It is different from the uniform covariance structure with two parameters as well as an unstructured multivariate normal distribution with $n_i(n_i - 1)/2$ parameters. Modeling the covariance structure in continuous time removes any requirement that the sequences or measurements on the different units are made at a common set of times.

Suppose there are five observations at times $0 < t_1 < t_2 < t_3 < t_4 < t_5$. Denote

$$a = \alpha^{(t_2^\lambda - t_1^\lambda)/\lambda}, \ b = \alpha^{(t_3^\lambda - t_2^\lambda)/\lambda}, \ c = \alpha^{(t_4^\lambda - t_3^\lambda)/\lambda}, \ d = \alpha^{(t_5^\lambda - t_4^\lambda)/\lambda}.$$

Consequently, the matrix can be written as

$$\Sigma = \sigma^2 \begin{bmatrix} 1 & a & ab & abc & abcd \\ a & 1 & b & bc & bcd \\ ab & b & 1 & c & cd \\ abc & bc & c & 1 & d \\ abcd & bcd & cd & d & 1 \end{bmatrix} \tag{3.21}$$

and the inverse of this covariance matrix is

$$\Sigma^{-1} = \frac{1}{\sigma^2} \begin{bmatrix} \frac{1}{1-a^2} & \frac{-a}{1-q^2} & 0 & 0 & 0 \\ \frac{-1}{1-a^2} & \frac{1-a^2b^2}{(1-a^2)(1-b^2)} & \frac{-b}{1-b^2} & 0 & 0 \\ 0 & \frac{-b}{1-b^2} & \frac{1-b^2c^2}{(1-b^2)(1-c^2)} & \frac{-c}{1-c^2} & 0 \\ 0 & 0 & \frac{-c}{1-c^2} & \frac{1-c^2d^2}{(1-c^2)(1-d^2)} & \frac{-d}{1-d^2} \\ 0 & 0 & 0 & \frac{-d}{1-d^2} & \frac{1}{1-d^2} \end{bmatrix}. \tag{3.22}$$

The elements of the covariance matrix are

$$\sigma^2(\Sigma^{-1})_{11} = [1 - \alpha^{2(t_2^\lambda - t_1^\lambda)/\lambda}]^{-1},$$

$$\sigma^2(\Sigma^{-1})_{kk} = [1 - \alpha^{2(t_k^\lambda - t_{k-1}^\lambda)/\lambda}]^{-1},$$

$$\sigma^2(\Sigma^{-1})_{j,j+1} = -[1 - \alpha^{2(t_{j+1}^\lambda - t_j^\lambda)/\lambda}]^{-1} \alpha^{(t_{j+1}^\lambda - t_j^\lambda)/\lambda}, \ 1 \leq j \leq k-1,$$

$$\sigma^2(\Sigma^{-1})_{jj} = \{[1 - \alpha^{2(t_j^\lambda - t_{j-1}^\lambda)/\lambda}][1 - \alpha^{2(t_{j+1}^\lambda - t_j^\lambda)/\lambda}]\}^{-1}[1 - \alpha^{2(t_{j+1}^\lambda - t_{j-1}^\lambda)/\lambda}], \ 1 < j < k,$$

$$\sigma^2(\Sigma^{-1})_{j1} = 0, \ |j-1| > 1.$$

In the case that variances are different, we may write the more general form for the covariance matrix, $\Sigma = A^{1/2}RA^{1/2}$, where $A = \mathrm{diag}(\sigma_k)$, $k = 1,\ldots,n_i$, and R is the correlation matrix.

We can also consider the damped exponential correlation structure here. Muñoz et al. (1992) introduced this structure. The model can handle slowly decaying autocorrelation dependence and autocorrelation dependence that decay faster than the commonly used first-order autoregressive model as well. In addition, the covariance structure allows for nonequidistant and unbalanced observations, thus efficiently accommodate the occurrence of missing observation.

Let $Y_i = (y_{i1},\ldots,y_{in_i})^T$ be a $n_i \times 1$ vector of responses at n_i time points for the ith individual ($i = 1,\ldots,N$). The covariate measurements X_i is an $n_i \times p$ matrix. Denote the n_i-vector s_i the times elapsed from baseline to follow-up with $s_{i1} = 0$, $s_{i2} = $ time from baseline to first follow-up visit on subject i; $s_{i,n_i} = $ time from baseline to last follow-up visit for subject i. The follow-up time can be scaled to keep s_i small positive integers of a size comparable to $\max_i\{n_i\}$ so that we can avoid exponentiation with unnecessarily large numbers. We assume that the marginal density on the ith subject, $i = 1,\ldots,N$, is

$$Y_i \sim \mathrm{MVN}(X_i\beta, \sigma^2 V_i(\alpha,\theta; s_i)), \quad 0 \le \alpha < 1, \quad \theta \ge 0; \tag{3.23}$$

and the (j,k) ($j < k$) element of V_i is

$$corr(Y_{ij}, Y_{ik}) = [V_i(\alpha,\theta; s_i)]_{jk} = \alpha^{(s_{ik}-s_{ij})^\theta}, \tag{3.24}$$

where α denotes the correlation between observations separated by one s-unit in time; θ is the "scale parameter" which permits attenuation or acceleration of the exponential decay of the autocorrelation function defining an AR(1). As attenuation is the most common in practical applications, we refer to this model as the damped exponential (DE). Given that most longitudinal data exhibit a positive correlation, it is sensible to limit α within nonnegative values.

For nonnegative α, the correlation structure given by (3.24) produces a variety of correlation structures upon fixing the scale parameter θ. Let I_B be the indicator function of the set B. If $\theta = 0$, then $corr(Y_{it}, Y_{i,t+s}) = I_{|s=0|} + \alpha I_{|s>0|}$, which is compound symmetry model; If $\theta = 1$, then $corr(Y_{it}, Y_{i,t+s}) = \alpha^{|s|}$, yielding AR(1); And as $\theta \to \infty$, $corr(Y_{it}, Y_{i,t+s}) \to I_{|s=0|} + \alpha I_{|s=1|}$, yielding MA(1); If $0 < \theta < 1$, we obtain a family of correlation structures with decay rate between those of compound symmetry and AR(1) models; For $\theta > 1$, it is a correlation structure with a decay rate faster than that of AR(1).

As we know, any correlation matrix should be positive definite with all nonnegative eigenvalues. This means that the correlation parameters are constrained. For example, for AR(1) model, we must have $|\alpha| \le 1$; for exchangeable model, we must have $1 \ge \alpha \ge -1/(n_i - 1)$.

For the unstructured structure, it is less clear how these parameters α_j should be constrained. Zhang et al. (2015) proposed an unconstrained parameterization for any correlation matrix using hyperspherical coordinates (the angles are free parameters between 0 and π).

Exercise 10   For the exchangeable correlation structure R with correlation parameter α and dimension n, show $|R| = (1-\alpha)^{n-1}\{1+(n-1)\alpha\}$, and

$$R^{-1} = \frac{1}{1-\alpha}\left(I - \frac{\alpha}{1-\alpha+\alpha n}\mathbf{ee}^{\top}\right) \tag{3.25}$$

in which $\mathbf{e}$ is a unit vector consist of 1s, $(1,1,...,1)$. The eigen-values are

$$\frac{(1-\rho)(1+(n-1)\alpha}{(1+(n-2)\alpha\}}.$$

Exercise 11   (Graybill Theorem 8.15.4) If R has the MA(1) correlation structure with the correlation parameter α, show

$$|R| = \prod_{j=1}^{n}\left\{1 + 2\alpha\cos(\frac{j\pi}{n+1})\right\}. \tag{3.26}$$

The eigenvalues are $1 + 2\alpha\cos(j/(n+1)\pi)$ for $j = 1,2,...,n$, and the maximum $\alpha \leq \min\{1, 1/(2\cos(\pi/(n+1)))\}$. The (j,k)-th element $(j \geq k)$ of its inverse R^{-1} is given by

$$b_{jk} = \frac{(1-b^{2n-2k+2})(b^{j+k+1} - b^{k-j+1})}{\alpha(1-b^2)(1-b^{2n+2})}, \tag{3.27}$$

where $b = (\sqrt{1-4\alpha^2}-1)/(2\alpha)$, i.e., $\alpha = -b/(1+b^2)$.

It is crucial to know the constraints so that we can make sure the resultant R matrix is meaningful when the parameters are estimated from the data. More discussion will be given towards the end of the next Chapter.

Davidian and Giltinan (1995) provided a comprehensive description on the mixed effect models and their computational algorithms when different variance function and covariance structures are used.

3.8   Random Effects Models

We briefly introduce the widely used random-effects model in longitudinal data analysis. More details will be given in Chapter 9.

The random-effects model can be regarded as a fully-parametric and distributional approach. The widely used linear model assumes $\mathbf{E}(Y_i|X_i) = X_i\beta$, i.e., we can decompose the observed vector Y_i as two parts: deterministic part and noise ϵ_i,

$$Y_i = X_i\beta + \epsilon_i,$$

in which $\mathbf{E}(\epsilon_i|X_i) = 0$.

As an alternative to directly modeling the correlations or the covariance of Y_i, we can decompose/model ϵ_i further as $Z_ib_i + \eta_i$, where b_i is referred to as the random effects associated with individual i. As we shall see, the presence of the same b_i vector in all observations from subject i induces correlations among these observations. This random-effects model explicitly identifies individual effects. In contrast to full multivariate models, which are not able to fit unbalanced data, the random effect model can handle the unbalanced situation. A model contains both fixed effects and random effects is often referred to as the mixed-effects model (not to be confused with the mixture model).

For multivariate normal data, the random-effects model can be described by two steps (Laird and Ware, 1982):

Step 1. For the ith experiment unit, $i = 1,\ldots,N$,

$$Y_i = X_i\beta + Z_i b_i + \eta_i, \tag{3.28}$$

where

X_i is a $n_i \times p$ "design matrix";

β is a $p \times 1$ vector of parameters referred to as fixed effects;

Z_i is a $n_i \times k$ "design matrix" that characterizes random variation in the response attributable to among-unit sources;

b_i is a $k \times 1$ vector of unknown random effects;

-, and η_i is distributed as $N(0,\mathcal{R}_i)$. Here $\mathcal{R}_i$ is an $n_i \times n_i$ positive-definite covariance matrix reflecting "measurement" errors. In practice, $\mathcal{R}_i$ is often taken as a diagonal matrix.

Step 2. β is considered fixed parameters at the population level, and b_i is also the unknown parameters but for subject i only. Here η_i is often assumed to be independent. The b_i values are distributed as $N(0,G)$, independently of each other and of the η_i. Here G is a $k \times k$ positive-definite covariance matrix.

The regression parameter vector β is the fixed effects, which are assumed to be the same for all individuals and have population-averaged interpretation. In contrast to β, the vector b_i is comprised of subject-specific regression coefficients.

The conditional mean of Y_i, given b_i, is

$$E(Y_i|b_i) = X_i\beta + Z_i b_i,$$

which is the ith subject's mean response profile. The marginal or population-averaged mean of Y_i is

$$E(Y_i) = X_i\beta.$$

Similarly,

$$\text{var}(Y_i|b_i) = \text{var}(\eta_i) = \mathcal{R}_i$$

and

$$\text{var}(Y_i) = \text{var}(Z_i b_i) + \text{var}(\eta_i) = Z_i G Z_i^\top + \mathcal{R}_i.$$

Thus, the introduction of random effects, b_i, induces correlation (marginally) among the Y_i. That is,

$$\text{var}(Y_i) = \Sigma_i = Z_i G Z_i^\top + \mathcal{R}_i,$$

which has nonzero off-diagonal elements. Based on the assumption on b_i and η_i, we have

$$Y_i \sim N_{n_i}(X_i\beta, \Sigma_i).$$

Chapter 4

Parameter Estimation

A longitudinal study is characterized by repeated measures from individuals through time. Analysis of longitudinal data plays an important role in many research areas, such as medical and biological research, economics, and finance as well. This leads to interesting statistical research on methods to take account of possible correlations among the repeated observations.

Let us first consider the independent data. For independent data, we only have two types of parameters to estimate, namely, regression parameters, variance parameters including the scale parameter. In most research literature, when count data is analyzed, the Poisson model is often used with $\text{Var}(y) = \phi \mathbf{E}(y) = \phi \mu$. However, the real variance structures may be very different from the Poisson model. There are at least two possible generalizations to the Poisson variance model,

(1). $V(\mu) = \mu^{\gamma}, 1 \leq \gamma \leq 2$;

(2). $V(\mu) = \gamma_1 \mu + \gamma_2 \mu^2$,

where $\gamma, (\gamma_1, \gamma_2)$ are unknown constants to be estimated. In previous chapter, we have considered the variance function $V(\mu) = \mu^{\gamma}$ and the corresponding likelihood function.

Independent data can be classified into two types: univariate observations and multivariate observations. For both of them, regression parameters can be estimated by GLM approach; for the later one, if it is a special case of longitudinal data, then GEE approach can also be employed. We use Gaussian, Quasi-likelihood, and other approaches to estimate variance parameters, which we will introduce later on.

For longitudinal data, we may proceed with the independent data analysis pretending the data were independent. This approach, in general, results in consistent β parameter estimators, but the efficiency can be low (Wang and Carey, 2003; Fitzmaurice, 1995), in other words, the standard errors of other estimators (incorporating within-subject correlations) can be made much smaller. Once a consistent estimator of β is available, the resultant residuals are valid for further analysis, i.e., we can then estimate and carry out inferences on the variance functions and correlation structures based on these residuals. Of course, β estimates can then be updated when better covariance matrices for Y_i are available.

DOI: 10.1201/9781315153636-4

4.1 Likelihood Approach

We first introduce how β can be estimated when ignoring the within-subject correlations. It is simple to estimate the regression parameter by adopting GLM approach when the independent data is univariate. Consider the univariate observations y_i, $i = 1,\ldots,N$ and $p \times 1$ covariate vector X_i. Let β be a $p \times 1$ vector of regression parameter and linear predictor $\eta_i = X_i^\top \beta$. Suppose $Y = (y_1,\ldots,y_N)$ follows a distribution from a specific exponential family as given by (3.2),

$$f(y;\theta,\phi) = \exp\left\{\frac{y\theta - b(\theta)}{\phi} + c(y,\phi)\right\}$$

with canonical parameter θ and dispersion parameter ϕ. For each y_i, the log-likelihood is

$$L_i(\beta,\phi) = \log f(y_i;\theta_i,\phi).$$

For $y_1,\ldots,y_N$, the joint log-likelihood is

$$L(\beta,\phi) = \sum_{i=1}^{N} \log f(y_i;\theta_i,\phi) = \sum_{i=1}^{N} L_i(\beta,\phi).$$

The score estimation function for β_j, $j = 1,\ldots,p$, can be derived by applying chain rule,

$$\frac{\partial L(\beta,\phi)}{\partial \beta_j} = \sum_{i=1}^{N} \frac{\partial L_i(\beta,\phi)}{\partial \beta_j}$$

$$= \sum_{i=1}^{N} \left\{\frac{y_i - \mu_i}{\phi_i} \frac{1}{\mathbf{V}(\mu_i)} \frac{\partial \mu_i}{\partial \eta_i} x_{ij}\right\}.$$

The estimating functions for β can be written as

$$U(\beta;\phi) = \sum_{i=1}^{N} D_i^\top V_i^{-1} S_i,$$

in which V_i is the variance function, and $S_i = y_i - \mu_i$. The MLE can be obtained by solving $U(\beta;\phi) = 0$. Usually, we assume ϕ, which is a constant for all observations, or for subject i or observation i, ϕ can be replaced by ϕ/m_i, where m_i $(i = 1,2,\ldots,N)$ are known weights. In these cases, ϕ can be removed from the estimating equations $U(\beta;\phi) = 0$, and it does not contribute towards estimation of β. However, ϕ does affect the variance of β estimator as an over-dispersion parameter.

If we assume the likelihood $L(\beta,\phi)$ is the true likelihood that generates the data, we can rely on likelihood-based inferences. For example, the covariance of the resultant $\hat{\beta}$ can be approximated by the inverse of the information matrix. Otherwise, asymptotic results can be derived from the estimating functions $U(\beta;\phi)$. An excellent reference on misspecified likelihood inference is White (1982).

4.2 Quasi-likelihood Approach

Wedderburn (1974) defined the quasi-likelihood, Q for an observation y with mean μ and variance function $V(\mu)$ by the equation

$$Q(y;\mu) = \int_y^\mu \frac{y-u}{V(u)} du \tag{4.1}$$

plus some function of y only, or equivalently by

$$\partial Q(y;\mu)/\partial\mu = (y-\mu)/V(\mu). \tag{4.2}$$

The deviance function, which measures the discrepancy between the observation and its expected value, is obtained from the analog of the log-likelihood-ratio statistic

$$D(y;\mu) = -2\{Q(y;\mu) - Q(y;y)\} = -2\int_y^\mu \frac{y-u}{V(u)} du. \tag{4.3}$$

The QL estimating functions are given by

$$\sum_{i=1}^N D_i^\top V_i^{-1}(Y_i - \mu_i) = 0, \tag{4.4}$$

where $D_i = (\partial\mu_i/\partial\beta^\top)_{n_i \times p}$ and V_i is the variance function.

If we are only concerned about parameter estimation, we can rely on estimating functions taking forms similar to (4.4) or other functions we constructed, and there is no need to worry about the likelihood functions because the known variance function is sufficient and ready to be incorporated (as weighting) in our estimation.

Under mild conditions, the Wedderburn form of QL can be used as a valid 'likelihood' function for estimation and inference and compare different linear predictors or different link functions on the same data. It cannot, however, be used to compare different variance functions on the same data. To this end, Nelder and Pregibon (1987) proposed extended-likelihood definition (EQL),

$$Q^+(y;\mu) = -\frac{1}{2}\log\{2\pi\phi\mathbf{V}(y)\} - \frac{1}{2}D(y;\mu)/\phi, \tag{4.5}$$

where $D(y,\mu)$ is the deviance as defined in (4.3), and ϕ is the dispersion parameter, $\mathbf{V}(y)$ is the variance function applied to the observation. When there exists a distribution of the exponential family with a given variance function, it turns out that the EQL is the saddle point approximation to that distribution. Thus Q^+, like Q, does not make a full distributional assumption but only the first two moments. The Quasi-likelihood and extended quasi-likelihood for a single observation y_{it} are listed in Table 4.1.

A distribution can be formed from an extended quasi-likelihood by normalizing $\exp(Q^+)$ with a suitable factor to make the sum or integral equal to unity. However, Nelder and Pregibon (1987) argued that the solution of the maximum quasi-likelihood equations would be little affected by the omission of the normalizing factor because it was often found that the normalizing factor changed rather little with those parameters.

Table 4.1 *Quasi-likelihood and extended quasi-likelihood for a single observation y_{it}. ϕ is the dispersion parameter. The extended quasi-likelihood is $Q^+(\mu_{it}, y_{it}) = -0.5\{\phi^{-1}D(\mu_{it}, y_{it}) + \log(2\pi\phi A(\mu_{it}))\}$. Note that $D(\mu_{it}, y_{it})$ and hence $Q^+(\mu_{it}, y_{it})$ differ from those in Table 9.1 of McCullagh and Nelder (1989).*

Variance $A(\mu_{it})$	Quasi-likelihood $Q^*(\mu_{it}, y_{it})$	Deviance $D(\mu_{it}, y_{it})$	Canonical Link $\eta_{it} = g(\mu_{it})$
1	$-\frac{(y_{it}-\mu_{it})^2}{2\phi}$	$(y_{it}-\mu_{it})^2$	μ_{it}
μ_{it}	$(y_{it}\log(\mu_{it})-\mu_{it})/\phi$	$y_{it}\log(\frac{y_{it}}{\mu_{it}})-(y_{it}-\mu_{it})$	$\log(\mu_{it})$
μ_{it}^2	$-(\frac{y_{it}}{\mu_{it}}+\log(\mu_{it}))/\phi$	$\frac{y_{it}}{\mu_{it}}-\log(\frac{y_{it}}{\mu_{it}})-1$	$-1/\mu_{it}$
μ_{it}^ζ	$\frac{\mu_{it}^{-\zeta}}{\phi}\{\frac{\mu_{it}y_{it}}{1-\zeta}-\frac{\mu_{it}^2}{2-\zeta}\}$	$\frac{y_{it}^{2-\zeta}-(2-\zeta)y_{it}\mu_{it}^{1-\zeta}+(1-\zeta)\mu_{it}^{2-\zeta}}{(1-\zeta)(2-\zeta)}$	$1/\{(1-\zeta)\mu_{it}^{\zeta-1}\}$
$\mu_{it}(1-\mu_{it})$	$y_{it}\log\{\frac{\mu_{it}}{1-\mu_{it}}\}+\log(1-\mu_{it})$	$-2\{y_{it}\log\{\frac{\mu_{it}}{1-\mu_{it}}\}+\log(1-\mu_{it})\}$	$\log\{\mu_{it}/(1-\mu_{it})\}$
$\mu_{it}+\frac{\mu_{it}^2}{k}$	$\{y_{it}\log(\frac{\mu_{it}}{k+\mu_{it}})+k\log(\frac{k}{k+\mu_{it}})\}$	$-2[y_{it}\log\{\frac{\mu_{it}(k+y_{it})}{y_{it}(k+\mu_{it})}\}+k\log(\frac{k+y_{it}}{k+\mu_{it}})]$	$\log(\frac{\mu_{it}}{k+\mu_{it}})$

Remark 1. It is not clear how to improve this deviance function by incorporating a correlation structure.

If we assume independence, the sum of deviances for all observations in $\mathcal{D}$ is

$$D(\beta; I, \mathcal{D}) = \sum_{i=1}^{N}\sum_{t=1}^{n_i} D(y_{it}; \mu_{it}). \tag{4.6}$$

Pan (2001b) derived the Akaike Information Criterion (AIC) based on the independence QL for model selection in longitudinal data analysis.

Later on, Hin and Wang (2009) discovered that this criterion is only valid for selecting useful ones among the predictors $(x_1, x_2, ..., x_p)$, but not for correlation structures because it is derived assuming the independence model.

For estimating β, the QL estimating functions are again given by

$$\sum_{i=1}^{N} D_i^\top V_i^{-1}(Y_i - \mu_i) = 0, \tag{4.7}$$

where V_i is a diagonal variance matrix here.

As for the parameters in V_i and ϕ, the extended QL becomes useful: an iterative approach between (4.7) and updating V_i can be adopted.

Note that $Q(y_{it}, y_{it})$ should be zero according to the definition. However, the $Q(y_{it}, y_{it})$ in Table 9.1 of McCullagh and Nelder (1989) does not become zero generally. For example, when $A(\mu_{it}) = \mu_{it}^2$, a constant $Q^*(y_{it}, y_{it}) = -\{1+\log(y_{it})\}/\phi$, is missing. Therefore, $Q^*(y_{it}, y_{it})$ needs to be subtracted in calculating $D(\beta; I, \mathcal{D})$ (Wang and Hin, 2009). Therefore, the quasi-likelihood outlined in Table 9.1 of McCullagh and Nelder (1989) should be used with care. To be more specific, we should define $Q(y_{it}, \mu_{it}) = Q^*(y_{it}, \mu_{it}) - Q^*(y_{it}, y_{it})$ so that $Q(y_{it}, y_{it}) = 0$.

In Section 3.2, we also defined the QL when y_i is a correlated multivariate vector. Regardless the integral in $Q(\mu; Y)$ is path-dependent or not, their derivatives with respect to β are unique,

$$U(\beta) = \sum_{i=1}^{N} D_i^\top V_i^{-1}(Y_i - \mu_i) = 0, \qquad (4.8)$$

which takes the same form as (4.7), but V_i now is a covariance matrix incorporating a hypothesized correlation matrix.

As for the asymptotic variance of the estimators, due to the lack of the true likelihood for further inference, we can rely on the approximating covariance of $U(\beta)$. We will provide more details when introducing the sandwich estimator. The book Heyde (1997) has provided a comprehensive description of the quasi-likelihood theory and inferences for martingale families.

4.3 Gaussian Approach

For the independence case we assume $n_i = 1$ for simplicity, the Gaussian likelihood given by (3.6), we have the score functions for the mean parameters β and all the other variance function parameters (denoted as τ including ϕ) are,

$$\partial L_G(\beta, \tau)/\partial \beta = \sum_{i=1}^{N} D_i^\top \frac{y_i - \mu_i}{\sigma_i^2} + \sum_{i=1}^{N} U_{\sigma_i}(\beta),$$

$$\partial L_G(\beta, \tau)/\partial \tau = \sum_{i=1}^{N} \sigma_i^{-2}\left(1 - \frac{(y_i - \mu_i)^2}{\sigma_i^2}\right)\frac{\partial \sigma_i^2}{\partial \tau},$$

where

$$U_{\sigma_i}(\beta) = \sigma_i^{-2}\left(1 - \frac{(y_i - \mu_i)^2}{\sigma_i^2}\right)\frac{\partial \sigma_i^2}{\partial \beta^\top}.$$

For example, if we let $g(\mu) = \mu^\gamma$, where $\gamma > 0$, the above likelihood function does not belong to the linear exponential distribution form because y_i and other power functions y_i are interacting with μ_i. If both mean and variance functions are correct, we have $\mathbf{E}\{\partial L_G(\beta, \tau)/\partial \beta\} = 0$. In ordinary regression, we only rely on the linear combinations of $S_i = (y_i - \mu_i)$ to gain protection against misspecification of the variance modeling. We can also achieve this by ignoring the second term $U_{\sigma_i}(\beta)$ in $\partial L_G(\beta, \tau)/\partial \beta$, and use only the first term for estimating β (with some notation abuse in U_G),

$$U_G(\beta;\tau) = \sum_{i=1}^{N} D_i^\top \frac{y_i - \mu_i}{\sigma_i^2}, \tag{4.9}$$

$$U_G(\tau;\beta) = \sum_{i=1}^{N} \sigma_i^{-2} \left\{ 1 - \frac{(y_i - \mu_i)^2}{\sigma_i^2} \right\} \frac{\partial \sigma_i^2}{\partial \tau^\top}. \tag{4.10}$$

This is precisely the approach of the iterative reweighted least-squares approach by iterating between (4.9) (updating β estimates) and (4.10) for updating σ_i weights. This makes one wonder if there is an underlying quasi-likelihood that produces such simpler estimating functions. We used "quasi" here because we know the "likelihood" we hypothesized is usually different to the true likelihood.

The Gaussian score function, obtained by differentiating equation (3.7) with respect to θ, for each component β_j in β is

$$u(\beta_j;\tau) = \partial L_G / \partial \beta_j$$

$$= \sum_{i=1}^{N} \left\{ \frac{\partial \mu_i}{\partial \beta_j} \Sigma_i^{-1} (Y_i - \mu_i) \right\}$$

$$+ \text{tr} \left\{ \sum_{i=1}^{N} \left[\Sigma_i^{-1} (Y_i - \mu_i)(Y_i - \mu_i)^T - I \right] \Sigma_i^{-1} (\partial \Sigma_i / \partial \beta_j) \right\}, \tag{4.11}$$

and for the variance parameter τ_j,

$$u(\tau_j;\beta) = \partial L_G / \partial \tau_j$$

$$= \text{tr} \left\{ \sum_{i=1}^{N} \left[\Sigma_i^{-1} (Y_i - \mu_i)(Y_i - \mu_i)^T - I \right] \Sigma_i^{-1} (\partial \Sigma_i / \partial \tau_j) \right\}. \tag{4.12}$$

A key condition for consistency of the estimator is that the estimating equations should be unbiased, at least asymptotically.

Again, to have protection to misspecification of $\Sigma_i = \phi V_i$, we will drop the second term in $u(\beta_j;\tau)$ and obtain the matrix form for β,

$$U(\beta;\tau) = \sum_{i=1}^{N} D^\top V_i^{-1} (Y_i - \mu_i), \tag{4.13}$$

$$u(\tau_j;\beta) = \text{tr} \left\{ \sum_{i=1}^{N} \left[\Sigma_i^{-1} (Y_i - \mu_i)(Y_i - \mu_i)^T - I \right] \Sigma_i^{-1} (\partial \Sigma_i / \partial \tau_j) \right\}. \tag{4.14}$$

We have $\mathbf{E}\{U(\beta;\tau)\} = 0$ when the mean assumption holds and $\mathbf{E}\{u(\tau_j;\beta)\} = 0$ when the covariance assumption holds. This will lead to consistency of β estimates even Σ_i matrices are incorrectly modeled.

Another perspective of looking at these above estimating functions is using the "decoupling" idea. If we write the mean μ_i explicitly as a function of β while write Σ_i explicitly as a function of β_* (as if it is a new set of parameters),

$$L_{G*}(\beta; \tau) = \sum_{i=1}^{N} \{\log\{|\Sigma_i(\beta_*)| + (Y_i - \mu_i(\beta))^\top \Sigma_i(\beta_*)^{-1}(Y_i - \mu_i(\beta))\}.$$

Minimization of $L_{G*}(\beta; \tau)$ with respect to β will lead to estimating functions given by (4.13).

Remark 2. $\log\{|\Sigma_i(\beta_*)|$ is deemed as a constant in this decoupling approach and it can be ignored.

Remark 3. In practice, one can simply replace β_* by the previous β estimates when minimizing L_{G*}. For Gaussian estimation, under mild regularity conditions, when the working correlation matrix is either correctly specified as the true one $\tilde{R}_i$ or chosen as the identity matrix (the independence model), the Gaussian estimators of regression and variance parameters are consistent (Wang and Zhao, 2007). To see this, we will check when we have $\mathbf{E}\{U(\beta; \tau)\} = 0$ and $\mathbf{E}\{u(\tau_j; \beta)\} = 0$. From equations (4.11) and (4.12), the unbiasedness condition for θ_j is

$$\mathbf{E}\left\{\mathrm{tr}\left[\Sigma_i^{-1}(Y_i - \mu_i)(Y_i - \mu_i)^T \Sigma_i^{-1} \frac{\partial \Sigma_i}{\partial \theta_j}\right]\right\} - \mathbf{E}\left\{\mathrm{tr}\left[\Sigma_i^{-1} \frac{\partial \Sigma_i}{\partial \theta_j}\right]\right\} = 0. \qquad (4.15)$$

Now we make some transformations of (4.15) to see the condition more clearly. For notation simplicity, let $\tilde{\Sigma}_i$ be the true covariance, thus $\tilde{\Sigma}_i = \mathbf{E}[(Y_i - \mu_i)(Y_i - \mu_i)^T] = A_i^{1/2} \tilde{R}_i A_i^{1/2}$, where $\tilde{R}_i$ is the true correlation structure.
The left-hand side of (4.15):

$$\mathbf{E}\left\{\mathrm{tr}\left[\Sigma_i^{-1}(Y_i - \mu_i)(Y_i - \mu_i)^T \Sigma_i^{-1} \frac{\partial \Sigma_i}{\partial \theta_j}\right]\right\} - \mathbf{E}\left\{\mathrm{tr}\left[\Sigma_i^{-1} \frac{\partial \Sigma_i}{\partial \theta_j}\right]\right\}$$

$$= -\mathrm{tr}\left[\frac{\partial \Sigma_i^{-1}}{\partial \theta_j} \tilde{\Sigma}_i\right] + \mathrm{tr}\left[\frac{\partial \Sigma_i^{-1}}{\partial \theta_j} \Sigma_i\right]$$

$$= -2\mathrm{tr}\left[\frac{\partial A_i^{-1/2}}{\partial \theta_j} R_i^{-1} A_i^{-1/2} A_i^{1/2} \tilde{R}_i A_i^{1/2}\right] + 2\mathrm{tr}\left[\frac{\partial A_i^{-1/2}}{\partial \theta_j} A_i^{1/2}\right]$$

$$= -2\mathrm{tr}\left[\frac{\partial A_i^{-1/2}}{\partial \theta_j} A_i^{1/2} R_i^{-1} \tilde{R}_i\right] + 2\mathrm{tr}\left[\frac{\partial A_i^{-1/2}}{\partial \theta_j} A_i^{1/2}\right]$$

$$= -2\mathrm{tr}\left[\frac{\partial A_i^{-1/2}}{\partial \theta_j} A_i^{1/2}(R_i^{-1}\tilde{R}_i - I)\right]. \qquad (4.16)$$

It is clearly that (4.16) will be 0 if $R_i = \tilde{R}_i$. As both the $\frac{\partial A_i^{-1/2}}{\partial \theta_j}$ and the A_i are diagonal matrices, (4.16) will be also 0 if the diagonal elements of $\{R_i^{-1}\tilde{R}_i - I\}$ are all 0. This will happen when $R_i = I$ because the diagonal elements of $\tilde{R}_i$ are all 1. Thus, we can conclude that under one of the two conditions: $R_i = \tilde{R}_i$ and $R_i = I$, the Gaussian estimation will be consistent.

This implies that we can use independent correlation structure if we have no idea about the true one, and the resulting estimator will be consistent under mild regularity conditions.

In general, (3.7) can be referred to as a "working" likelihood function to provide a sensible solution to supplying the working parameters needed for the mean parameter estimation.

In longitudinal data analysis, the mean response is usually modeled as a function of time and other covariates. Profile analysis (assuming each time point is a category instead of a continuous variable) and parametric curves are the two popular strategies for modeling the time trend. In a parametric approach, we model the mean as an explicit function of time. If the profile means appear to change linearly over time, we can fit linear model over time; if the profile means appear to change over time in a quadratic manner, we can fit the quadratic model over time. Appropriate tests may be used to check which model is the better choice. Clearly, profile analysis is only sensible if each subject has the same set of observation times.

4.4   Generalized Estimating Equations (GEE)

The main objective in longitudinal studies is to describe the marginal expectation of the outcome as a function of the predictor variables or covariates. As repeated observations are made on each subject, correlation among a subject's measurement may be generated. Thus the correlation should be accounted for to obtain an appropriate statistical analysis. However, the GLM only handles independent data.

When the outcome variable is approximately Gaussian with a constant variance, linear models are often used to fit the data or after a power transformation. But for non-Gaussian longitudinal data, few models are available in the literature. The limited number of models for non-Gaussian longitudinal data is partly due to the lack of a rich class of multivariate distributions such as the multivariate Gaussian for the repeated observations for each subject. Due to this reason, likelihood-based methods are generally not popular.

Even for the covariance modeling, it is usually like a bitter pill we have to swallow if we have to assume the true covariance is a member of the hypothesized family, especially when the subjects are recruited from different areas and over an extended study period. Irregular observations times and missing data makes us more brave to impose such assumptions.

To this end, the generalized estimating equations (GEE) was developed by Liang and Zeger (1986). The framework of GEE is based on quasi-likelihood theory.

In addition, a "working" correlation matrix for the repeated observations for each subject is nominated in GEE. We denote the "working" correlation matrix by $R_i(\alpha)$, which is a matrix with unknown parameters α. The nominated "working" correlation matrix to be correct to produce consistent β estimators.

4.4.1   Estimation of Mean Parameters β

Recall that we assume the data are collected from N subjects indexed by subscript i. The outcome vector for subject i consists of n_i observations, $Y_i = (y_{i1}, \ldots, y_{in_i})^\top$, and covariates (predictors, features) are recorded in matrix X_i, of dimension $n_i \times p$. The first column of X_i is usually a vector of 1s so that the intercept term is included unless specified. The corresponding observation times for Y_i are in the vector $T_i = (t_{ij})$, of dimension n_i.

The marginal mean of Y_i is μ_i, and the fundamental assumption is the parametric relationship between X_i and $\mathbf{E}(Y_i)$,

$$\mu_i = h(X_i^\top \beta),$$

where β is a $p \times 1$ vector of parameters.

Akin to the GLM framework, we also need to specify the mean function in relation to the covariates. Apart from choosing systematic components and the distribution family for the responses, we can also establish such mean function $h(X_i, \beta)$ as in nonlinear regression. For distributional assumptions, we can select normal, gamma, inverse Gaussian random components for continuous data and binary, and multinomial, Poisson components for discrete data.

Remark 4.   The marginal mean μ_i depends on the p predictors via a single index $\theta_i = X_i^\top \beta$. This assumption can be easily relaxed to be more general,

$$\mu_i = h(X_i, \beta).$$

This mean model will include nonlinear regression setup and multiple index models as long as h is given a priori and the dimension of β, p, is fixed and usually much smaller than n.

Remark 5.   If a spline function is assumed to model h, the number of parameters, p, will expand as n or n_i increases. This will lead to a new area of semiparametric modeling for longitudinal data analysis (Lin and Ying, 2001; Lin and Carroll, 2006; Li et al., 2010; Hua and Zhang, 2012).

The inverse of h or h itself is often referred to as the "link" function. In quasi-likelihood, the variance of Y_{ij}, σ_{ij}^2 is expressed as a known function g of the expectation μ_{ij}, i.e.,

$$\sigma_{ij}^2 = \phi g(\mu_{ij}),$$

where ϕ is a scale parameter. Write $\mathrm{Cov}(Y_i|X_i)$ as $\phi A_i^{1/2} \tilde{R}_i(\alpha) A_i^{1/2}$, where $\tilde{R}_i(\alpha)$ is the true correlation matrix and $A_i = \mathrm{diag}(\sigma_{ij})$ is the diagonal variance matrix. Here ϕ is regarded as an overdispersion parameter in GLM setting. The univariate specification is $\mathbf{E}(y_{ij}) = \mu_{ij}$ and $\mathrm{var}(y_{ij}) = \phi \sigma_{ij}^2$. The mean model is a familiar generalized linear model (GLM) specification, and the covariance factor V_i with elements (σ_{ij}) is the GLM variance function or a generalized version with additional parameters to be more flexible.

The working covariance matrix for Y_i is given by ϕV_i, where

$$V_i = A_i^{1/2} R_i(\alpha) A_i^{1/2}, \tag{4.17}$$

where A_i is an $n_i \times n_i$ diagonal matrix with var(y_{ij}) as the jth diagonal element.

Remark 6. The variance σ_{ij} and hence A_i^2 are parameterized by γ, which includes ϕ.

Remark 7. The correlation matrix R_i is parameterized by α.

Based on quasi-likelihood and the set up of "working" correlation matrix, Liang and Zeger(1986) derived the generalized estimating equations that produce consistent estimators of the regression coefficients under mild regularity conditions. The variances of the β estimators can also be estimated by the so-called sandwich estimator consistently without requiring the V_i be correctly modeled. Bear in mind, the covariance matrix of the β estimates depends on both the chosen working matrix and the true covariance matrix.

The "working independence" model has been advocated for a variety of robustness properties (Pepe and Anderson, 1994).

Sutradhar and Das (1999) concluded on the basis of limited asymptotic studies with a fixed balanced design that if there is a substantial risk of misspecification of the working correlation model, the analyst may be better off with the independence working model. In contrast to this, it has been suggested that efficiency gains can be obtained provided misspecification is not "too bad". Note that independence models have been found to be fully efficient only when the covariate pattern is the same for all individuals (Lipsitz et al., 1994; Fitzmaurice, 1995; Mancl and Leroux, 1996). Fitzmaurice (1995) also considered more realistic cases when the within-cluster covariate differs among clusters and found that independence models suffer substantial losses in efficiency. Mancl and Leroux (1996) provided a thorough study on independence model when the true correlation structure is exchangeable.

Suppose τ collects all the variance parameters including the overdispersion parameter ϕ. For a given covariance model, ϕV_i, the generalized estimating equations (GEE) can be expressed as,

$$U_{\text{GEE}}(\beta; \tau, \alpha) = \sum_{i=1}^{N} D_i^\top V_i^{-1}(Y_i - \mu_i) = (0)_{p \times 1}, \qquad (4.18)$$

where $\mu_i = (\mu_{i1}, \ldots, \mu_{in_i})^\top$ and $D_i = \partial \mu_i / \partial \beta^\top$. In order to solve β from the above equations, both D_i and V_i must be given up to the parameter vector β, and all other parameters such as α and γ (except for ϕ) must be supplied a priori.

For convenience, we will let U_i be the contribution of subject i in U_{GEE}, i.e., $U_i = D_i^\top V_i^{-1} S_i$, where $S_i = Y_i - \mu_i$. The above GEE is not a stranger to us. Its form is the same as the estimating functions derived from the linear exponential family (4.1). In particular, when V_i is diagonal (independence working correlation model), U_{GEE} becomes the estimating functions derived from the QL approach (4.4), or the general V_i corresponds to the multivariate version given by (4.8). The decoupled version from Gaussian estimation also takes the form, see (4.13).

Suppose that $\hat{\alpha}$ and $\hat{\gamma}$ are $\sqrt{N}$-consistent estimators obtained somehow via other approaches, the GEE (4.18) above will solve $U_{\text{GEE}}(\beta; \hat{\alpha}, \hat{\gamma}) = 0_{p \times 1}$. In general, $\hat{\alpha}$ and $\hat{\gamma}$ will also require a $\sqrt{N}$-consistent estimator of β so that valid residuals can be obtained for α and γ estimators. So, the GEE estimator of β, $\hat{\beta}_{\text{GEE}}$ is essentially obtained by solving $U_{GEE}(\beta; \hat{\alpha}(\beta), \hat{\gamma}(\beta)) = 0_{p \times 1}$. We will omit the subscript indicting the dimension when there is no confusion.

Under mild regularity conditions and prerequisite that the link function is correctly specified under minimal assumptions about the time dependence, Liang and Zeger (1986) showed that as $N \longrightarrow \infty$, $\hat{\beta}_{GEE}$ is a consistent estimator of β and that $\sqrt{N}(\hat{\beta}_R - \beta)$ is asymptotically multivariate Gaussian with covariance matrix given by

$$\lim_{N \to \infty} N \left(\sum_{i=1}^{N} D_i^\top V_i^{-1} D_i \right)^{-1} \left[\sum_{i=1}^{N} D_i^\top V_i^{-1} Cov(Y_i) V_i^{-1} D_i \right] \left(\sum_{i=1}^{N} D_i^\top V_i^{-1} D_i \right)^{-1} . \quad (4.19)$$

This matrix can be estimated consistently without any knowledge on $Cov(Y_i)$ directly because $Cov(Y_i)$ can be simply replaced by the residual product $\hat{S}_i \hat{S}_i^\top$ and α, β and ϕ by their estimates in equation (4.20). This leads to the well-celebrated sandwich estimator for the covariance of $\hat{\beta}_{GEE}$,

$$\widehat{V_{GEE}} = \left(\sum_{i=1}^{N} D_i^\top V_i^{-1} D_i \right)^{-1} \left[\sum_{i=1}^{N} D_i^\top V_i^{-1} \hat{S}_i \hat{S}_i^\top V_i^{-1} D_i \right] \left(\sum_{i=1}^{N} D_i^\top V_i^{-1} D_i \right)^{-1} , \quad (4.20)$$

in which all the matrices, D_i, V_i if needed, are evaluated at the final parameter estimates, $\hat{\beta}_{GEE}$, $\hat{\alpha}$, and $\hat{\gamma}$.

When R_i is correctly specified and α is known, the generalized estimating equations given by (4.18) is optimal for β in the class of linear functions of Y_i according to the Gauss-Markov theorem. In practice, the correlation matrix is often unknown and we rely on the estimation of α given β. This leads to simultaneous or iterative estimation for β and α (Zhao and Prentice, 1990; Liang et al., 1992).

A key robustness property motivating the widespread use of generalized estimating equations is the consistency of the solution $\hat{\beta}$ whether or not the working correlation structure R_i is correctly specified. Careful modeling of R_i will improve the efficiency of $\hat{\beta}$, especially when the sample size is not large (Albert and McShane, 1995). The optimal asymptotic efficiency of $\hat{\beta}$ is obtained when R_i coincides with the true correlation matrix. In practice, R_i would be chosen to be the most reasonable for the data based on either statistical criteria or biological background.

Although the GEE approach can provide consistent regression coefficient estimators, the estimation efficiency may fluctuate greatly according to the specification of the "working" covariance matrix. The "working" covariance has two parts: one is the "working" correlation structure; the other is the variance function. The existing literature has been focused on the specification of the "working" correlation while the variance function is often assumed to be correctly chosen, such as Poisson variance and Gaussian variance function. In real data analysis, if the variance function is also misspecified, the estimation efficiency will be low. It is, therefore, crucial to model the variance function in order to improve the estimation efficiency of β. Relevant work is abundant (Dempster (1972); Davidian and Carroll (1987); O'Hara-Hines (1998); Wang and Lin (2005); Wang and Hin (2009)).

The GEE approach estimates β using only linear combinations of the data as given by (4.18). In some cases, the variance and covariance can contain information on μ_{ij} and hence β, and it then becomes sensible to construct estimating functions based on the quadratic functions of the data. GEE2 proposed by Prentice and Zhao (1991) aims to construct the "optimal" combination of both linear and quadratic residuals for estimating β. Prentice and Zhao (1991) and Fitzmaurice et al. (1993) also obtained this type of estimating functions using a likelihood-based method. The GEE2 can, therefore, be regarded as driven by absorbing higher moment information. The connection between the likelihood equations and the GEE2 is well described by Fitzmaurice et al. (1993).

It is true that the GEE2 may produce more efficient estimators of β when the third and fourth moments can be correctly specified. However, as pointed out by Prentice and Zhao (1991), models for means and covariances can typically be sensibly specified and are readily interpreted. Models for higher order parameters are usually specified in an *ad hoc* fashion and are unlikely to accurately approximate the real complexity.

Apart from computationally intensive as n_i gets large, GEE2 approach also has the following drawbacks: (i) no consistency of the mean parameter estimates under covariance misspecification; (ii) no efficiency will be gained if the third or the fourth moment functions are incorrect (Liang et al. 1992). In general, little can be gained in incorporating the third and fourth moments in the estimation. Therefore, in practice, one may wish to restrict the assumptions on the mean and covariance functions only.

4.4.2   Estimation of Variance Parameters τ

As we can see, the GEE approach requires input of a covariance matrix for each subject. We now present how the variance function can be estimated, and then proceed to the correlation matrix in §4.4.3. Recall τ collects all the variance parameters including the overdispersion parameter ϕ.

In the GEE approach, much attention is put on the specification of the working correlation structure while ignoring the importance of modeling the variance function.

The existing approaches for estimating the variance parameters in correlated data analysis do not "decouple" the variance and correlation in the sense that the estimator of the variance parameters always rely on the assumed correlation structure such as the maximum likelihood (ML) and the restricted maximum likelihood (REML) (Davidian and Giltinan, 1995; Pinheiro and Bates, 2000).

4.4.2.1   Gaussian Estimation

Suppose R_i is the nominated correlation matrix for estimating γ. Note that this choice does not have to be the same as the "working" correlation matrix. The Gaussian log-likelihood for the complete data vector Y is

$$\log \prod_{i=1}^{N} \left[\{\det(2\pi\Sigma_i)\}^{-1/2} \exp\left\{ -\frac{1}{2}(Y_i - \mu_i)\Sigma_i^{-1}(Y_i - \mu_i)^\top \right\} \right],$$

where $\Sigma_i = \phi V_i$, the working covariance matrix. The corresponding Gaussian score function with respect to τ relies on minimizing the Gaussian -2log-likelihood as shown by expression (4.12),

$$u(\tau_j; \beta) = \text{tr}\left[\sum_{i=1}^{N}\left\{\{\Sigma_i^{-1}(Y_i - \mu_i)(Y_i - \mu_i)^\top - I\}\Sigma_i^{-1}\frac{\partial \Sigma_i}{\partial \tau_j}\right\}\right].$$

According to in §4.3, for using the true $R = \tilde{R}_i$ or the naïve independence $R = I$, the Gaussian estimator of τ is consistent under mild conditions.

If the specified variance function is far from the true one, the correct choice of the working correlation matrix may no longer be the true one. In real data analysis, it is difficult to determine the variance function, and possibly, we will choose an incorrect variance function. Therefore, akin to the correlation parameters, the variance parameters are also subject to the pitfalls discussed by Crowder (1995) under model misspecification. Therefore, well-behaved estimators for those working parameters are highly desirable to avoid possible convergence problems and to result in efficient estimators of β (Wang and Carey, 2003). The Gaussian working likelihood should be a convenient and useful approach to analyzing longitudinal data.

4.4.2.2 Extended Quasi-likelihood

We can adopt the extended QL under independence for estimating the variance parameters. The variance function is $\mathbf{V}(\mu) = \mu^\gamma$, and $\text{Var}(y) = \phi \mathbf{V}(\mu)$. Under these settings, the quasi-likelihood contribution of a single observation is

$$Q_i^+(y_i; \mu_i, \gamma) = -\frac{1}{2}\log\{2\pi \mathbf{V}(y_i)\} - \frac{1}{2}D(y_i; \mu_i), \qquad (4.21)$$

where $D(y_i; \mu_i)$ is deviance function given by

$$D(y_i; \mu_i) = -2\int_{y_i}^{\mu_i}\frac{y_i - u}{\mathbf{V}(u)}du.$$

For the variance function $\mathbf{V}(\mu) = \mu^\gamma$, the deviance function is

$$D(y; \mu) = \begin{cases} 2\{y\log(y/\mu) - (y - \mu)\} & \gamma = 1, \\ 2\{y/\mu - \log(y/\mu) - 1\} & \gamma = 2, \\ \frac{2\{y^{2-\gamma} - (2-\gamma)y\mu^{1-\gamma} + (1-\gamma)\mu^{2-\gamma}\}}{(1-\gamma)(2-\gamma)} & \text{otherwise.} \end{cases}$$

When $\gamma \neq 1, 2$, the score estimation equation with respect to γ can be expressed by $\sum_{i=1}^{N}\sum_{j=1}^{n_i}q_{ij}(y_{ij}; \gamma) = 0$, where

$$q_{ij}(y_{ij}; \gamma) = \partial Q_{ij}^+/\partial \gamma$$

$$= -\frac{1}{2}\log y_{ij} - \frac{y_{ij}^{2-\gamma}(2\gamma - 3)}{(1-\gamma)^2(2-\gamma)^2} - \frac{y_{ij}^{2-\gamma}\log y_{ij}}{(1-\gamma)(2-\gamma)}$$

$$+ \frac{y_{ij}\mu_i^{1-\gamma}\log\mu_{ij}}{1-\gamma} - \frac{y_{ij}\mu_{ij}^{1-\gamma}}{(1-\gamma)^2} - \frac{\mu_{ij}^{2-\gamma}\log\mu_{ij}}{2-\gamma} + \frac{\mu_{ij}^{2-\gamma}}{(2-\gamma)^2}.$$

The success of the extended QL is nicely demonstrated by Nelder and Pregibon (1987).

4.4.2.3   Nonlinear Regression

Denote the residuals by $r_{ij} = Y_{ij} - \mu_{ij}$. Clearly, the variance information lies in r_i^2. The squared residuals r_{ij}^2 have expectation $\sigma_{ij}^2 = \phi g(\mu_{ij})$. This suggests a nonlinear regression on $\{r_{ij}^2\}$ versus $\phi g(\mu_{ij})$. The estimator $\hat{\tau}$ can be obtained by the least-squares approach (Davidian and Carroll, 1987),

$$\sum_{i=1}^{N} \{r_{ij}^2 - \phi g(\hat{\mu}_{ij})\}^2.$$

For normal data the squared residuals have approximate variance proportional to σ_{ij}^4; the generalized weighted least squares can be considered as well,

$$\sum_{i=1}^{N} \{r_i^2 - \phi g(\hat{\mu}_{ij})\}^2 / g^2(\hat{\mu}_{ij}),$$

where $\hat{\gamma}$ is a preliminary estimator for γ.

Alternatively, one can also apply the LS approach to the logarithmic transformed residuals,

$$\sum_{i=1}^{N} [2\log(|r_{ij}|) - \log\{\phi g(\hat{\mu}_{ij})\}]^2.$$

One pitfall is that if any of the residuals are close to 0, the regression could be adversely affected by these large negative outliers.

4.4.2.4   Estimation of Scale Parameter ϕ

In GLM setting, ϕ is regarded as the overdispersion parameter. In GEE approach, this scale parameter does not affect the β estimation, however, the variance of the β estimates is, in fact, proportional to ϕ. Therefore, the final ϕ estimate can play critical role in statistical inferences (hypothesis testing and confidence intervals). In fact, the scale parameter is one of the variance parameters. The Gaussian approach, the extended QL, the regression are all applicable for estimating ϕ.

All the fundamental approaches, such as the maximum likelihood, quasi-likelihood, Gaussian and moment, can be considered for estimating ϕ, depending on the overall longitudinal model used in the analysis. Moment estimation is a sensible and can be advantageous to the likelihood-based approaches. Moment estimation for ϕ is often based on the Pearson residuals,

$$\frac{1}{N-p} \sum_{i=1}^{N} \sum_{j=1}^{n_i} \frac{(y_{ij} - \hat{\mu}_{ij})^2}{g(\hat{\mu}_{ij})},$$

where p is the number of regression parameters.

For the Gaussian likelihood approach and given $\hat{\beta}$, we can write the independence log-likelihood as an explicit function of ϕ (Carroll and Ruppert, 1982),

$$l(\beta,\gamma,\phi) = -\frac{\log\phi}{2}\sum_{i=1}^{N}n_i - \frac{1}{2}\sum_{i=1}^{N}\sum_{j=1}^{n_i}\log\{g(\hat{\mu}_{ij})\} - (2\phi)^{-1}\sum_{i=1}^{N}\sum_{j=1}^{n_i}\{y_{ij}-\mu_{ij}\}^2/g(\hat{\mu}_{ij})$$

leading to

$$\hat{\phi}_{Gau} = \frac{1}{\sum_{i=1}^{N}n_i}\sum_{i=1}^{N}\sum_{j=1}^{n_i}(y_{ij}-\mu_{ij})^2/g(\hat{\mu}_{ij}).$$

4.4.3 Estimation of Correlation Parameters

The GEE approach is an elegant method for estimation of β, which can be regarded as the optimal linear combinations of the data according to the Gauss-Markov theorem (Heyde, 1997).

The working matrix of $\mathrm{Cov}(Y_i)$ by ϕV_i, which can also be written as form $\phi A_i^{1/2} R_i A_i^{1/2}$, with $A_i = \mathrm{diag}\{\mathrm{Var}(y_{it})\}$ and $R_i(\alpha)$ being the working correlation matrix of Y_i. Apart from the variance parameter τ, there are also "nuisance" parameters in R_i that we have to estimate. In the same way as the τ estimates are used in A_i, the α estimates for the correlation matrix will then be used in the GEE for estimating β.

For estimation of correlation parameters α, the estimation method often depends on the chosen correlation structure. Liang and Zeger (1986) illustrated these methods by several examples.

The number of correlation parameters and the estimator of α vary from case to case in the literature. Most researchers follow Liang and Zeger (1986), who discussed a number of important special cases. We now discuss how to model the correlation matrix. Commonly used or encountered correlation structures for R_i are stationary in the sense that the correlation matrix does not change when all observations are shifted in time.

4.4.3.1 Stationary Correlation Structures

The following structures are the typical "working" correlation structures, and the estimators are used to estimate the "working" correlations. The number of correlation parameter varies according to different "working" correlation structures as introduced in Section 3.7.

- *Independence model*

$$R = \begin{pmatrix} 1 & 0 & 0 & \dots & 0 \\ 0 & 1 & 0 & \dots & 0 \\ \dots & & & & \\ 0 & 0 & 0 & \dots & 1 \end{pmatrix}_{n\times n}.$$

- *Compound symmetry (exchangeable or equal correlation)*

$$R = \begin{pmatrix} 1 & \alpha & \alpha & \cdots & \alpha \\ \alpha & 1 & \alpha & \cdots & \alpha \\ \cdots & & & & \\ \alpha & \alpha & \alpha & \cdots & 1 \end{pmatrix}_{n \times n}.$$

- *First order autoregressive: AR(1)*

$$R = \begin{pmatrix} 1 & \alpha & \alpha^2 & \cdots & \alpha^{n-1} \\ \alpha & 1 & \alpha & \cdots & \alpha^{n-2} \\ \alpha^2 & \alpha & 1 & \cdots & \alpha^{n-3} \\ \cdots & & & & \\ \alpha^{n-1} & \alpha^{n-2} & \alpha^{n-3} & \cdots & 1 \end{pmatrix}_{n \times n}.$$

- *Moving average*

$$R = \begin{pmatrix} 1 & \alpha & 0 & \cdots & 0 & 0 \\ \alpha & 1 & \alpha & \cdots & 0 & 0 \\ 0 & \alpha & 1 & \cdots & 0 & 0 \\ \cdots & & & & & \\ 0 & 0 & o & \cdots & \alpha & 1 \end{pmatrix}_{n \times n}.$$

- *m-dependent correlation*
  This is a generalization of the above first-order moving average model to order m,

$$\text{Corr}(y_{ij}, y_{ij+t}) = \begin{cases} 1 & t = 0 \\ \alpha_t & t = 1, 2, \ldots, m \\ 0 & t > m \end{cases}.$$

- *Toeplitz*
  This is essentially a special case of m-dependent model with the largest m possible, $m = n$. Note that this structure is also "stationary" as the correlation is lag-dependent but not the starting time.

$$R = \begin{pmatrix} 1 & \alpha_1 & \alpha_2 & \cdots & \alpha_{n-1} \\ \alpha_1 & 1 & \alpha_1 & \cdots & \alpha_{n-2} \\ \alpha_2 & \alpha_1 & & \cdots & \alpha_{n-3} \\ \cdots & & & & \\ \alpha_{n-1} & \alpha_{n-2} & \alpha_{n-3} & \cdots & 1 \end{pmatrix}_{n \times n}.$$

The above correlation structures are only appropriate for equally-spaced measurements, and all subjects must have the same observation times in order to share the same correlation structure.

If stationary is in question, or observations are irregularly spaced, we may consider

- *Unstructured*

$$R = \begin{pmatrix} 1 & \alpha_{12} & \alpha_{13} & \cdots & \alpha_{1n} \\ \alpha_{21} & 1 & \alpha_{23} & \cdots & \alpha_{2n} \\ \alpha_{31} & \alpha_{32} & 1 & \cdots & \alpha_{3n} \\ \cdots & & & & \\ \alpha_{n1} & \alpha_{n2} & \alpha_{n3} & \cdots & 1 \end{pmatrix}.$$

Note that, again, all subjects must have the *same* observation times in order to share the same correlation structure.

All these correlation structures can be easily implemented in R package *nlme*. For example, if ID is the factor for the n subject levels, the generic R code for fitting the AR(1) covariance model by using *corr=corAR1*(form = 1—id) in *gls* function.

The moment estimators can be easily constructed for these models. The estimators are all expressed based on the Pearson residuals. Denote Pearson residual by $\hat{e}_{ij} = (y_{ij} - \mu_{ij})/A_{ij}$, $j = 1, 2, ..., n$. where A_{ij} is the jth diagonal element of the diagonal variance matrix A_i. Note that $\sigma_{ij}^2 = \phi A_{ij}$, and $\mathbf{E}(\hat{e}_{ij}^2) \approx \phi$.

For m-dependent model, we will use the only t-lagged residuals for estimating α_t, $t = 1, 2, ..., m$,

$$\hat{\alpha}_t = \frac{2 \sum_{i=1}^{N} \sum_{j \leq n_i - t} \hat{e}_{ij} \hat{e}_{ij+t}}{\sum_{i=1}^{N} \sum_{j \leq n_i - t} (\hat{e}_{ij}^2 + \hat{e}_{ij+t}^2)}. \tag{4.22}$$

This estimator (with $m = 1$) is also applicable to both moving average and the AR(1) model.

For exchangeable structure, we can make use of all the lagged residual products,

$$\hat{\alpha} = \frac{2 \sum_{i=1}^{N} \sum_{j \neq k} \hat{e}_{ij} \hat{e}_{ik}}{\sum_{i=1}^{N} \sum_{j \neq k} (\hat{e}_{ij}^2 + \hat{e}_{ik}^2)}.$$

For the unstructured correlation, we will use all the pairwise products, the estimator for the parameter α_{jk} will rely on the residuals at time j and time k,

$$\hat{\alpha}_{jk} = \frac{2 \sum_{i=1}^{N} \hat{e}_{ij} \hat{e}_{ik}}{\sum_{i=1}^{N} (\hat{e}_{ij}^2 + \hat{e}_{ik}^2)}.$$

Note that all these correlation estimators are bounded between -1 and 1 according to the Cauchy inequality. But this does not necessarily mean the resultant matrices are always positive definite. This brings another issue as pointed out by Crowder (1995). We will discuss this further towards the end of this section.

The correlation matrix $R(\alpha)$ has $q = \dim(\alpha)$ parameters. For example, in a study involving T equidistant follow-up visits, the unstructured correlation matrix for an individual with complete data will have $q = T(T-1)/2$ correlation parameters; if the repeated observations are assumed exchangeable, R will have the compound symmetry structure, and $q = 1$. A parsimonious parametrization of the correlation structure is desired in order to optimize the efficiency of the estimation procedure.

Note that an underlying assumption here is that the correlations depends on j, k but not on subject i. Sometimes, this may not hold that the covariance matrix varies between subjects. Even the unstructured matrix will not be able to accommodate this because the subjects cannot share the same correlation matrix anymore.

For observations at continuous times, and at irregular times, t_{i1}, t_{i2}, t_{i3}, and t_{i4}, for example, we may wish to consider the *First order autoregressive with continuous time: CAR(1)*,

$$
R = \begin{pmatrix}
1 & \alpha^{|t_{i1}-t_{i2}|} & \alpha^{|t_{i1}-t_{i3}|} & \cdots & \alpha^{|t_{i1}-t_{in}|} \\
\alpha^{|t_{i2}-t_{i1}|} & 1 & \alpha^{|t_{i2}-t_{i3}|} & \cdots & \alpha^{|t_{i2}-t_{in}|} \\
\alpha^{|t_{i3}-t_{i1}|} & \alpha^{|t_{i3}-t_{i2}|} & 1 & \cdots & \alpha^{|t_{i3}-t_{in}|} \\
\multicolumn{5}{c}{\cdots} \\
\alpha^{|t_{in}-t_{i1}|} & \alpha^{|t_{in}-t_{i2}|} & \alpha^{|t_{in}-t_{i3}|} & \cdots & 1
\end{pmatrix}_{n \times n} .
$$

In this case each subject will have a different correlation matrix depending on the specific observation times. Muñoz et al. (1992) proposed a damped exponential correlation structure for modeling multivariate outcomes.

The damped exponential correlation structure applies a power transformation on the lag, two observations with lag s is modeled as if the time lag is s^φ in CAR(1) model, the correlation is α^{s^φ}, and φ is a damping parameter. The correlation structures of compound symmetry, first-order autoregressive, and first-order moving average processes can be obtained by assuming $\varphi = 0$, $\varphi = 1$, and $\varphi \to \infty$.

Thus, the correlation structures with $q = 2$ can model quite parsimoniously a variety of forms of dependence, including slowing decaying autocorrelation functions and autocorrelation functions that decay faster than the commonly used first-order autoregressive model.

However, estimation of α is not straightforward as the moment estimators cannot be constructed easily. We will, therefore, need to construct supplementary estimating functions for α. Further complications may arise when sampling is involved in nested, spatial and temporal random effects. Intuitively, careful modeling the correlation structure should improve the efficiency of estimation. Diggle et al. (1994) provides a comprehensive review of relevant techniques.

4.4.3.2 Generalized Markov Correlation Structure

The vector of observation times for subject i is $(t_{i1}, \ldots, t_{i,n_i})$, The generalized Markov correlation structure assumes the Markov correlation structure after applying a Box-Cox type of power transformation on time. The correlation between Y_{ij} and Y_{ik} is given, for $j \neq k$, by $\alpha^{\Delta_i(j,k;\lambda)}$, where $\Delta_i(j,k;\lambda) = |t_{ij}^\lambda - t_{ik}^\lambda|/\lambda$ Núñez Anton and Woodworth (1994). As one can see, this correlation function is not stationary in time any more unless $\lambda = 1$. The generalized Markov correlation structure accommodates irregular and nonlattice-valued observation times, which is quite norm in practice (Shults and Chaganty, 1998). This correlation is quite rich and flexible and is appropriate in many practical cases. A case of particular interest is the genetic distance among family members when data are clustered/coordinatized in pedigrees (Trègouët et al., 1997).

In many cases, the correlation parameter describes the association between responses from the same cluster (such as family, parents, or area), and may be of scientific interest. For example, intraclass correlation is commonly used to measure the degree of similarity or resemblance siblings. In these cases, the efficient estimation of the association parameters would be valuable. In genetic studies, we often rely on associations for prediction. So we have to come up with a correlation structure and estimate the correlation-related parameters (correlations may be governed by covariates such as dam weight in developmental studies). Correlation and mean parameters will be equally important in these cases.

Apart from simple moment estimators, the supplementary set of estimating functions for α can be constructed more elegantly to enhance the estimation efficiency and avoid possible pitfalls (Lipsitz et al., 1991; Prentice and Zhao, 1991; Liang et al., 1992; Hall and Severini, 1998). In analysis of real data, misspecification is probably the norm and the efficiency of $\hat{\beta}$ depends on how close the working matrix is to the true correlation. Because $\hat{\beta}$ values obtained from (4.18) depend on the values of α used, estimation methods for α are, therefore, of importance for improving the efficiency of estimation of β. On the other hand, in many cases, estimates of α may be of scientific interest as well (Lipsitz et al. 1991). More detials will be given in Chapter 9.5.

4.4.3.3 Second Moment Method

In fact, the moment estimators can be formulated using a linear combination of the residual products, which reflects the association between any two responses from the same subject. It is, therefore, quite natural to consider the products of all pairwise residual products. Let $z_{ikj} = (y_{ik} - \mu_{ik})(y_{ij} - \mu_{ij})$, following the GEE idea for estimating β the following estimating functions for α,

$$U_M(\alpha;\beta) = \sum_{i=1}^{N} \frac{\partial \delta_i}{\partial \alpha} Q_i^{-1}(Z_i)(Z_i - \delta_i),$$

in which Z_i is the vector of residuals consisting of (z_{ikj}), $\delta_i = E(\mathbf{Z}_i)$ and Q_i is a working covariance matrix for Z_i. Note that the dimension of Z_i is $\binom{n_i}{2}$.

The working covariance Q_i may be chosen to be diagonal weighting. For binary responses, $\mathrm{var}(z_{ikj}) = \delta_{ikj}(1 - \delta_{kij})$. In general, if $\mathrm{var}(z_{ikj})$ is unknown, one may choose the diagonal elements of Q as $\sigma_{ik}^2 \sigma_{ij}^2$. Prentice and Zhao (1991) proposed a few specifications of Q to account for the correlations between high orders.

4.4.3.4 Gaussian Estimation

Whittle (1961) introduced Gaussian estimation, which uses a normal log-likelihood as a vehicle for estimation without assuming that the data are normally distributed. Crowder (1985) also proposed this approach for analyzing correlated binomial data. Later on, Crowder (2001) further proposed the modified Gaussian estimation for analysis of repeated measures. The modification is to decouple the β parameters in the covariance matrix from the β in the mean so that the estimating functions for β will ignore the information of β in V_i and V_i will play the role of weighting. In our current setting, we have additional parameters γ in A_i and α in R_i.

Let θ be the vector collecting all the parameters including β, and $\tau = (\gamma, \phi, \alpha)$, where γ and α are the parameter vectors in A_i and the correlation structure, respectively. When necessary, we will write μ_i and V_i explicitly as $\mu_i = \mu_i(\beta)$, $A_i = A_i(\beta, \gamma)$, $R_i = R_i(\alpha)$ and $V_i = V_i(\beta, \tau)$. We shall be interested in consistent α estimators that have protection against misspecified R_i to improve the asymptotic efficiency of the β estimators. The working likelihood -2log-likelihood for the data $(Y_1, ..., Y_N)$ is

$$L_G(\theta) = \sum_{i=1}^{N} [\log\{\det(2\pi\phi V_i)\} + \phi^{-1}(Y_i - \mu_i)V_i^{-1}(Y_i - \mu_i)^\top]. \qquad (4.23)$$

We can rely on Gaussian estimation to get consistent estimate of correlation parameters in the same way as for the variance parameters. The score function with respect to α_j has the same form as (4.12),

$$\text{tr}\left[\sum_{i=1}^{N} \{V_i^{-1}(Y_i - \mu_i)(Y_i - \mu_i)^\top - I_{n_i}\}V_i^{-1}(\partial V_i/\partial \alpha_j)\right]. \qquad (4.24)$$

Therefore, we know that when the covariance is correctly specified or when the independence model is used, the estimating functions for γ from the Gaussian method will be unbiased from 0. A safe estimation approach is to simply use the independent model when it is difficult to model the true correlations as the resulting estimators of (β, γ) will be consistent under mild regularity conditions. The full estimation procedure for $\theta = (\beta, \gamma, \phi, \alpha)$ can be described by the following iterative algorithm.

(i) obtain initial estimates of β from the independence model or some other naive model;

(ii) for given $\beta = \hat{\beta}$, apply the Gaussian approach using the independence model to obtain/update the estimator of γ, ie, minimizing $L_{G0}(\hat{\beta}, \phi, \gamma)$ with respect to (ϕ, γ), as also given by (3.6),

$$L_{G0}(\hat{\beta}, \phi, \gamma) = \sum_{i=1}^{N} [\log\{\det(\phi\hat{A}_i^2)\} + \phi^{-1}(Y_i - \hat{\mu}_i)^\top \hat{A}_i^{-2}(Y_i - \hat{\mu}_i)],$$

where $\hat{\mu}_i$ and $\hat{A}_i$ are evaluated at $\beta = \hat{\beta}$.

(iii) For given $(\beta, \gamma) = (\hat{\beta}, \hat{\gamma})$, apply the Gaussian-likelihood approach to the independence model to obtain/update the estimator of α, i.e., maximizing $L_G(\hat{\beta}, \hat{\gamma}, \hat{\phi}, \alpha)$ given by (3.7), with respect to α. Or alternatively, the estimator of α may be obtained by other existing methods (Wang and Carey, 2003);

(iv) For given $\tau = (\hat{\gamma}, \hat{\phi}, \hat{\alpha})$, update $\hat{\beta}$ by maximizing the Gaussian function $\hat{G}(\beta)$ with respect to β, where $\hat{G}(\beta)$ is $G(\beta, \hat{\gamma}, \hat{\phi}, \hat{\alpha})$ given by (4.23) with V_i being evaluated at the previous estimate $\hat{\theta}$;

(v) Iterate between (ii) and (iv) until desired convergence as in the ordinary.

For estimation of β, the β in V_i is decoupled from the β in μ_i. The resulting estimating equations for β, therefore, become the same as (4.18) which have consistency protection against misspecification of V_i. Recall that $e_i = A_i^{-1}(y_i - \mu_i)$. The corresponding estimating functions for (γ, ϕ, α) can be written as

$$g(\gamma_j; \theta) = \text{tr} \left[\sum_{i=1}^{N} (e_i e_i^\top - \phi I_{n_i}) A_i^{-1} (\partial A_i / \partial \gamma_j) \right], \tag{4.25}$$

$$\phi = \frac{1}{\sum_i n_i} \sum_{i=1}^{N} \sum_{j=1}^{n_i} (y_{ij} - \mu_{ij})^2 / v_{ij},$$

$$g(\alpha_j; \theta) = \text{tr} \{ \sum_{i=1}^{N} (e_i e_i^\top - \phi R_i)(\partial R_i^{-1} / \partial \alpha_j) \}. \tag{4.26}$$

Estimator of β, γ and α are obtained from the iterative method or the joint estimating functions, (4.18), together with (4.25) and (4.26).

4.4.3.5 Quasi Least-squares

Shults and Chaganty (1998) suggest the quasi-least squares (QLS) for parameter estimation, which unifies the estimation of mean and variance parameter estimation but the variance parameter estimates are biased even when the correlation structure is correctly specified. Their method is desirable only if the correlation parameters are of no interest but has to be used to take account for dependence in estimating the mean parameters.

In notation, the QLS estimates of α are obtained by minimizing $\sum_i (Y_i - \mu_i)^\top V_i^{-1} (Y_i - \mu_i)$.

The resulting estimating function for the parameter α_j in α is

$$U_Q(\alpha_j; \beta) = \sum_{i=1}^{N} \epsilon_i' \mathbf{P}_{ij} \epsilon_i,$$

in which $\mathbf{P}_{ij} = \partial \mathbf{R}_i^{-1} / \partial \alpha_j$, which can also be written as $\mathbf{P}_{ij} = -\mathbf{R}_i^{-1} \partial \mathbf{R}_i / \partial \alpha_j \mathbf{R}_i^{-1}$. Shults and Chaganty (1998) also found that compared with the ad hoc methods in the literature, it has smaller empirical risk of producing infeasible correlation parameter estimates and smaller mean square error in the estimates of β when the correlation is small or moderate. The bias in the QLS estimates can be easily removed according to Chaganty and Shults (1999). More details, extensions and interesting applications can be found in the book Shults and Hilbe (2014).

To avoid nonconvergence or the infeasibility problem pointed out by Crowder (1995), explicit expressions of $\hat\alpha$ that are constrained within the sensible region are attractive. This leads Chaganty (1997) to consider the QLS method.

The problem is the QLS estimators are inconsistent. The bias-corrected version of Chaganty and Shults (1999) for the AR(1) working model is

$$\hat\alpha_{QLS1} = \frac{\sum_k \sum_{|i-j|=1} \hat\epsilon_{ki} \hat\epsilon_{kj}}{\sum_k \left(\hat\epsilon_{k1}^2 + 2 \sum_{i=1}^{n} \hat\epsilon_{ki}^2 + \hat\epsilon_{kn}^2 \right)}. \tag{4.27}$$

Clearly, we have $\hat\alpha_{QLS1} \to \gamma = \alpha$, when the true correlation structure is either AR(1), exchangeable or MA(1) with parameter α.

If working matrix is exchangeable instead of AR(1), and the corresponding QLS estimate is $\hat{\alpha}_{QLS}$, the bias corrected estimate is obtained from

$$\sum_{i,j}(\alpha_{ij}) = \frac{\{1+(n-1)\hat{\alpha}_{QLS}\}^2}{1+(n-1)\hat{\alpha}_{QLS}^2}.$$

If the true correlation is exchangeable, we can verify that the limit of the bias corrected estimate of α as

$$\frac{2+(n-2)\hat{\alpha}_{QLS}}{1+(n-1)\hat{\alpha}_{QLS}^2}\hat{\alpha}_{QLS} = \alpha.$$

In order to obtain consistent estimates of α for other types of working matrices (exchangeable and MA(1)), Chaganty and Shults (1999) suggested to obtain an initial QLS estimate using AR(1) working model, $\hat{\alpha}_{QLS}$, and then adjust the estimates by assuming the correlation structure believed to be true is either exchangeable or MA(1). The resulting estimate of α, $\hat{\alpha}_{QLS1}$, has the same expression as (4.27). Note that one needs to choose the working correlation matrices twice, and they do not have to be of the same structure. In theory, the initial estimate $\hat{\alpha}_{QLS}$ can also be based on exchangeable, MA(1) models, or any other correlation structures. But the initial estimate from AR(1) model leads to the above simple expressions.

4.4.3.6   Conditional Residual Method

Carey et al. (1993) proposed using the alternating logistic regression based on conditional residuals to estimate α. For any pair (k, j), $k < j$, we let

$$\xi_{ikj} = \mu_{ik} + (\sigma_{ikj}/\sigma_{ijj})(y_{ij} - \mu_{ij}),$$

and $s_{ikj} = y_{ik} - \xi_{ikj}$. Note that ξ_{ikj} has a marginal mean of μ_{ik}, and in most cases such as multivariate normal, binary and multinomial responses ξ_{ikj} is exactly the conditional mean, $\mathbf{E}(y_{ik}|y_{ij})$. In general, ξ_{ikj} can be regarded as a linear approximation to the conditional mean $\mathbf{E}(y_{ik}|y_{ij})$. Their idea can be easily extended to general cases.

Let Z_k and ξ_k be the vectors (of length $\binom{n_i}{2}$) consisting of s_{ikj} and ξ_{ikj}, respectively. We have the following estimating equation for α,

$$U_{CR}(\alpha;\beta) = \sum_{i=1}^{N}\left(\frac{\partial \xi_i}{\partial \alpha}\right)^{\top} \text{diag}(\Gamma_i^{-1})W_{i=0}, \tag{4.28}$$

in which Γ_i is a working variance vector, which can be chosen as $\{\text{var}(y_{ik}|y_{ij})\}$. In general, Γ_i may have to be chosen as the identity matrix, unless some mean-variance relationship can be specified such as the case of binary responses. It is easily to verify that $\mathbf{E}(U_{CR}(\alpha))$ is unbiased from 0 if either Γ_k is the identity matrix (and ξ_{kij} does not have to be the true condition mean), or ξ_{ikj} is the true conditional mean (and Γ_i may depend on ξ_i).

For multiple binary responses, Carey et al. (1993) found that using conditional residuals produce much more efficient estimates of correlation parameters than using unconditional residuals. This was further demonstrated by Lipsitz and Fitzmaurice (1996).

4.4.3.7 Cholesky Decomposition

In the conditional residual method, the identity matrix may have to be used for Γ_i when the conditional variance is unknown, which is often the case. Therefore, each residual is treated equally. This may be reasonable if the correlation is the same between each pairwise observations. This motivates us to consider using all previous responses rather than each single individual previous response.

Wang and Carey (2004) proposed Cholesky decomposition method to improve the estimation efficiency and guarantee feasibility of solutions. The basic idea was first presented at the Eastern North American Region International Biometric Society (ENAR) 1999 meeting and also at the Biostatistics Department Seminar, Harvard School of Public Health.

Let $R_i^{-1} = C_i^\top C_i$ and C_i is a lower triangular matrix, which can be obtained from Cholskey decomposition. Consider the transformed variable $H_i = C_i A_i^{-1}(Y_i - \mu_i)$, which are unbiased from 0 and an identity covariance matrix. Further more, we decompose C_i into

$$
\begin{pmatrix}
c_{11} & 0 & 0 & \cdots & 0 \\
0 & c_{22} & 0 & \cdots & 0 \\
0 & 0 & c_{33} & \cdots & 0 \\
\cdots & & & & \\
0 & 0 & 0 & \cdots & c_{nn}
\end{pmatrix}
+
\begin{pmatrix}
0 & 0 & 0 & \cdots & 0 \\
c_{21} & 0 & 0 & \cdots & 0 \\
c_{31} & c_{32} & 0 & \cdots & 0 \\
\cdots & & & & \\
c_{n1} & c_{n2} & c_{n3} & \cdots & 0
\end{pmatrix}
= J_i + B_i.
$$

We can now write H_i as

$$
J_i A_i^{-1}(Y_i - \zeta_i),
$$

in which $\zeta_i = \mu_i - A_i J_i^{-1} B_i A_i^{-1}(Y_i - \mu_i)$, which is a linear conditional mean vector. The second term $A_i J_i^{-1} B_i A_i^{-1}(Y_i - \mu_i)$ adjusts the mean based on previous responses if correlations between responses exist.

Remark 8. Note that ζ_i are linear predictors using the first two moments and the previous observations (their residuals). Therefore, in case of correct specifying R_i, we would expect ζ_i be centered as 0 and orthogonal to each other.

It is easily to show that

$$
\mathbf{E}\left(\frac{\partial \zeta_i}{\partial \beta}\right) = A_i J_i^{-1} C_i A_i^{-1} D_i,
$$

and

$$
\begin{aligned}
\mathrm{Cov}(Y_i - \zeta_i) &= \mathrm{Cov}\{A_i(I_i + J_i^{-1} B_i) A_i^{-1}(Y_i - \mu_i)\} \\
&= A_i J_i^{-1}(J_i + B_i) R_i (J_i + B_i)^\top J_i^{-1} A_i \\
&= A_i^2 J_i^{-2}.
\end{aligned}
$$

We will denote the diagonal matrix, $J_i^2 A_i^{-2}$, as W_i, for convenience. It is, therefore, sensible to consider the following estimating functions for α,

$$
U_{\mathrm{Chol}}(\alpha; \beta) = \sum_{i=1}^{N} \left(\frac{\partial \zeta_i}{\partial \alpha}\right)^\top W_i(Y_i - \zeta_i), \tag{4.29}
$$

in which W_i is a weighting diagonal matrix. In fact, for any chosen diagonal matrix W_i that is independent of the data, the estimating functions $U_{\mathrm{Chol}}(\alpha)$ given by (4.29) is unbiased from zero.

Remark 9. *It is easy to show* $\mathbf{E}(U_{\mathrm{Chol}}(\alpha;\beta)) = 0$.

We first rewrite $Y_i - \zeta_i$ as $A_i(I_i + J_i^{-1} B_i)A_i^{-1}(Y_i - \mu_i)$, and hence for any component α, α_i,

$$\frac{\partial \zeta_k}{\partial \alpha_i} = A_i F_{ik} A_i^{-1}(Y_i - \mu_i).$$

The estimating function $U_{\mathrm{Chol}}(\alpha;\beta)$ can, therefore, be written in the following quadratic form

$$\sum_{i=1}^{N}(Y_i - \mu_i)^{\top} A_i^{-1} M_{ik} A_i^{-1}(Y_i - \mu_i), \tag{4.30}$$

where $M_{ik} = F_{ik}^{\top} A_i^2 W_i^{-1}(I + J_i^{-1} B_i)$ and $F_{ik} = \partial(J_i^{-1} B_i)/\partial \alpha_k$.

We also have

$$\mathbf{E}\{(Y_i - \mu_i)^{\top} A_i^{-1} M_{ik} A_i^{-1}(Y_i - \mu_i)\} = \mathrm{tr}\left\{M_{ik}\mathrm{Cov}(A_i^{-1}(Y_i - \mu_i))\right\}$$
$$= \mathrm{tr}\left\{F_{ik}^{\top} A_i^2 W_i^{-1}(I + J_i^{-1} B_i)R_i\right\}$$
$$= \mathrm{tr}\left\{F_{ik}^{\top} A_i^2 W_i^{-1} J_i^{-1}(J_i + B_i^{\top})^{-1}\right\}$$
$$= \mathrm{tr}(L_k F_{ik}),$$

in which $L_i = C_i^{-1} J_i^{-1} W_i^{-1} A_i^2$ is a lower triangle matrix. Because F_{ik} is also a lower triangle matrix with all leading elements being 0, we have $\mathrm{tr}(L_i F_{ik}) = 0$, which completes the proof.

When C_i and B_i are taken to be an upper triangular matrix (this is equivalent to reverse the time order), we can also obtain a similar version of $U_{\mathrm{Chol}}(\alpha;\beta)$. In general, these two sets are very similar. The average or sum can be used as suggested by Wang and Carey (2004). This will also symmetrize the residual appearance as we will see later in AR(1) model. An advantage of the use of $\mathbf{U}_{Chol}$ over $\mathbf{U}_G$ is that U_{Chol} is free from the scale parameter ϕ.

After some algebra, we can rewrite (4.30) as

$$U_{\mathrm{Chol}}(\alpha;\beta) = \sum_{i=1}^{N} \mathbf{E}\left(\frac{\partial \zeta_i}{\partial \beta}\right)^{\top} W_i(Y_i - \zeta_i). \tag{4.31}$$

So the same "working" matrix are be used in $U(\beta)$ and $U_{\mathrm{Chol}}(\alpha;\beta)$. Note that there is an expectation operator, "E", in (4.31) so that quadratic terms are eliminated. A similar relationship was found by Wang (1999a) between conditional (transitional) models and marginal models.

If $\mathbf{E}(y_{ik}|y_{i1}, \cdots, y_{ik-1}) = \zeta_i$ is a linear function of past responses, and $\mathbf{V}_i^*$ is the corresponding conditional variance matrix (diagonal), the transitional or conditional model would rely on

$$U_c(\beta) = \sum_{i=1}^{N}\left(\frac{\partial \zeta_i}{\partial \beta}\right)^{\top} V_i^{*-1}(Y_i - \zeta_i),$$

for estimating β. Note that $\mathbf{E}(Y_i - \zeta_i)$ are unbiased from 0 even in the true conditional means are not linear functions. To ensure $\mathbf{E}\{U_c(\beta)\} = 0$ and hence robustness to mis-specification of conditional means, we may wish to replace the Jacobian matrix and the conditional variance matrix with their expectations.

Remark 10. ζ_i are linear combinations of past responses only, and can, therefore, be interpreted as some linear predictions. As the covariance matrix of $Y_i - \zeta_i$ is diagonal, namely, $A_i^2 J_i^{-2}$, the components can be roughly regarded as independent.

Remark 11. In the case of $W_i = J_i^2 A_i^{-2}$, as we suggested, $M_{ik} = F_{ik}^{\top} J_i C_i$, where $F_{ik} = \dfrac{\partial J_i^{-1} B_i}{\partial \alpha_k}$.

We now consider two widely used correlation structures, the first-order autoregression (AR(1)) and the equicorrelation structures. The first one is often used in longitudinal studies, and the latter is often used to account for cluster settings (Lipsitz and Fitzmaurice, 1996).

For the AR(1) model with correlation parameter α, the (i,j)th element of R is $\alpha^{|j-i|}$, $1 \le i, j \le n$, and $\det(R) = (1-\alpha^2)^{n-1}$. We have

$$\mathbf{R}^{-1} = \frac{1}{1-\alpha^2} \begin{pmatrix} 1 & -\alpha & 0 & 0 \\ -\alpha & 1+\alpha^2 & -\alpha & 0 \\ \vdots & & & \\ 0 & 0 & -\alpha & 1 \end{pmatrix}.$$

The Cholskey decomposition of $\mathbf{R}^{-1}$ leads to

$$C = \frac{1}{\sqrt{1-\alpha^2}} \begin{pmatrix} \sqrt{1-\alpha^2} & 0 & 0 & \cdots & 0 & 0 \\ -\alpha & 1 & 0 & \cdots & 0 & 0 \\ 0 & -\alpha & 1 & \cdots & 0 & 0 \\ \vdots & \vdots & \vdots & \vdots & \vdots & \vdots \\ 0 & 0 & 0 & \cdots & -\alpha & 1 \end{pmatrix},$$

$$J^{-1}\mathbf{B} = \begin{pmatrix} 0 & 0 & 0 & \cdots & 0 & 0 \\ -\alpha & 0 & 0 & \cdots & 0 & 0 \\ 0 & -\alpha & 0 & \cdots & 0 & 0 \\ \vdots & \vdots & \vdots & \vdots & \vdots & \vdots \\ 0 & 0 & 0 & \cdots & -\alpha & 0 \end{pmatrix} \quad \text{and} \quad J^2 = \frac{1}{1-\alpha^2}\text{diag}\begin{pmatrix} 1-\alpha^2 \\ 1 \\ 1 \\ \vdots \\ 1 \end{pmatrix}.$$

We, therefore, have for the ith subject,

$$\zeta_i = \mu_i + \alpha_i \begin{pmatrix} 0 \\ \sigma_{i2}/\sigma_{i1}\epsilon_{i1} \\ \sigma_{i3}/\sigma_{i2}\epsilon_{i2} \\ \vdots \\ \sigma_{in}/\sigma_{in-1}\epsilon_{in-1} \end{pmatrix},$$

in which $\epsilon_{ik} = y_{ik} - \mu_{ik}$ and α_k is the correlation parameter which may depend on the subject-specific covariates through parameters α. The contribution of the ith subject to $U_{\text{Chol}}(\alpha;\beta)$, given by (4.29) is

$$\frac{1}{1-\alpha_i^2}\sum_{k=2}^{n}\left(\frac{\epsilon_{ik}}{\sigma_{ik}}-\frac{\epsilon_{ik-1}}{\sigma_{ik-1}}\alpha_i\right)\frac{\epsilon_{ik-1}}{\sigma_{ik-1}}.$$

The resulting estimate of α from $U_{\text{Chol}}(\alpha;\beta)$ is

$$\hat{\alpha}_{\text{Chol}}=\frac{\sum_{i=1}^{N}\sum_{k=2}^{n_i}\hat{\epsilon}_{ik}\hat{\epsilon}_{jk}}{\sum_{i=1}^{N}\sum_{k=1}^{n_i-1}\hat{\epsilon}_{ik}^2},$$

in which $\hat{\epsilon}_{ik}=\epsilon_{ik}/\sigma_{ik}$.

If R is the equicorrelated (exchangeable correlation) matrix with the constant correlation parameter α for subject i, we have $\det(R)=(1-\alpha_i)^{n-1}\{1+(n-1)\alpha_i\}$, and

$$R^{-1}=\frac{1}{1-\alpha}\left(\mathbf{I}-\frac{\alpha}{1-\alpha+\alpha n}\mathbf{1}\right)$$

in which $\mathbf{1}$ is a unit matrix of $n\times n$. After some algebra, we have $R^{-1}=C^TC=(J+B^T)(J+B)$, in which

$$J^2=\text{diag}\begin{pmatrix}\dfrac{1}{(1-\alpha^2)^{-1}}\\ \vdots\\ \dfrac{1+(n-2)\alpha}{1+(n-2)\alpha-(n-1)\alpha^2}\end{pmatrix},$$

and

$$J^{-1}\mathbf{B}=\text{diag}\begin{pmatrix}1\\ -\alpha\\ \vdots\\ \dfrac{-\alpha}{1+(n-2)\alpha}\end{pmatrix}\begin{pmatrix}0 & 0 & 0 & \cdots & 0 & 0\\ 1 & 0 & 0 & \cdots & 0 & 0\\ 1 & 1 & 0 & \cdots & 0 & 0\\ \vdots & & & & & \\ 1 & 1 & 1 & \cdots & 1 & 0\end{pmatrix}.$$

Therefore,

$$\zeta_i=\mu_i+\alpha\begin{pmatrix}0\\ \dfrac{\sigma_{i2}}{\sigma_{i1}}\epsilon_{i1}\\ \vdots\\ \dfrac{\sigma_n}{1+(n-2)\alpha}\sum_{j=1}^{n-1}\dfrac{\epsilon_{ij}}{\sigma_{ij}}\end{pmatrix}.$$

The contribution of the i-th subject to the estimating function is

$$\sum_{k=2}^{n}\left\{a_k^{-1}\left(\sum_{j=1}^{k-1}\frac{\epsilon_{ij}}{\sigma_{ij}}\right)\left(\frac{\epsilon_{ik}}{\sigma_{ik}}-\frac{\alpha}{1+(k-2)\alpha}\sum_{j=1}^{k-1}\frac{\epsilon_{ij}}{\sigma_{ij}}\right)\right\},$$

in which $a_k=(1-\alpha)\{1+(k-1)\alpha\}\{1+(k-2)\alpha\}$. It is easily to see that the above expression has 0 expectation under the equicorrelation assumption.

For the general Markov correlation, the corresponding estimating functions for λ are Wang and Carey (2004),

$$U_{\text{Chol}}(\lambda;\beta) = \sum_{i=1}^{N} \sum_{j=2}^{n_i} \frac{l_{ij}\alpha^{d_{ij}}}{1-\alpha^{2d_{ij}}}\{(\alpha^{d_{ij}}\epsilon_{ij-1} - \epsilon_{ij})\epsilon_{ij-1} + (\alpha^{d_{ij}}\epsilon_{ij} - \epsilon_{ij-1})\epsilon_{ij}\},$$

$$U_G(\lambda;\beta) = \mathbf{U}_{\text{Chol}}(\lambda;\beta) + \sum_{i=1}^{N} \sum_{j=2}^{n_i} \frac{l_{ij}\alpha^{2d_{ij}}}{(1-\alpha^{2d_{ij}})^2}\{(\epsilon_{ij} - \alpha^{d_{ij}}\epsilon_{ij-1})^2 + (\epsilon_{ij-1} - \alpha^{d_{ij}}\epsilon_{ij})^2$$
$$- 2\phi(1-\alpha^{2d_{ij}})\},$$

in which

$$l_{ij} = \partial d_{ij}/\partial\lambda = \frac{t_{ij}^{\lambda}\log(t_{ij}) - t_{ij-1}^{\lambda}\log(t_{ij-1})}{\lambda} - d_{ij}/\lambda.$$

The special case of AR(1) model ($\lambda = 1$), $t_i = i$ and $n_i = 3$ for all subjects. The solution to $U_{\text{Chol}}(\alpha)$ is

$$\hat{\alpha}_{\text{Chol}} = \frac{\sum_{i=1}^{N}(\epsilon_{i1}\epsilon_{i2} + \epsilon_{i2}\epsilon_{i3})/2}{\sum_{i=1}^{N}(\epsilon_{i1}^2 + 2\epsilon_{i2}^2 + \epsilon_{i3}^2)/4}. \tag{4.32}$$

Another advantage of $\hat{\alpha}_{\text{Chol}}$ is that it always lies in the sensible range of $(-1, 1)$.

4.4.4 Covariance Matrix of $\hat{\beta}$

Recall that the GEE estimator $\hat{\beta}$ of β by solving the following estimating equations,

$$U_{\text{GEE}}(\beta;\tau,\alpha) = \sum_{i=1}^{N} D_i^{\top} V_i^{-1}(Y_i - \mu_i) = 0. \tag{4.33}$$

Suppose the true parameters are denoted as $\tilde{\beta}$. The covariance of $U(\tilde{\beta})$ can be approximated by $E\{U_{\text{GEE}}(\tilde{\beta};\tau,\alpha)U_{\text{GEE}}(\tilde{\beta};\tau,\alpha)^{\top}\} = \sum_{i=1}^{N} D_i^{\top} V_i^{-1}\tilde{\Sigma}_i V_i^{-1} D_i$.

Suppose we have constructed $U(\alpha,\tau;\beta)$ for all the variance and correlation parameters as supplementary estimating functions to $U_{\text{GEE}}(\beta;\tau,\alpha)$ as given by (4.18), the final estimates of (β,α,τ) from the joint estimating functions

$$U(\theta) = \begin{pmatrix} U_{\text{GEE}}(\beta;\tau) \\ U(\alpha,\tau;\beta) \end{pmatrix}.$$

Using the method of delta approximation, we can approximate the covariance of the estimates as

$$\Delta^{-1}\text{Cov}(U(\theta))(\Delta^{\top})^{-1},$$

in which Δ is the Jacobian matrix $(\partial U(\theta)/\partial\theta)$ evaluated at the estimated values. The covariance of $U(\theta)$ can be estimated from the products of the subject residuals (contributions to the estimating functions).

For the proposed method, Δ can be replaced by

$$
\begin{pmatrix}
\sum_{i=1}^{N} D_i^{\top} V_i^{-1} D_i & \mathbf{0} \\
\mathbf{E}\{\partial U(\alpha, \tau; \beta)/\partial \beta^{\top}\} & \mathbf{E}\{\partial U(\alpha, \tau; \beta)/\partial(\alpha, \tau)\}
\end{pmatrix}.
$$

Under mild regularity conditions, $\hat{\beta}_R$ is consistent as $N \longrightarrow \infty$, and $\sqrt{N}(\hat{\beta}_R - \beta)$ is asymptotically multivariate Gaussian with covariance matrix V_R given by

$$
V_R = \lim_{N \to \infty} N \left(\sum_{i=1}^{N} D_i^{\top} V_i^{-1} D_i \right)^{-1} \left[\sum_{i=1}^{N} D_i^{\top} V_i^{-1} \mathrm{Cov}(Y_i) V_i^{-1} D_i \right] \left(\sum_{i=1}^{N} D_i^{\top} V_i^{-1} D_i \right)^{-1}.
$$

The covariance matrix of $\hat{\beta}$ can be consistently estimated by the sandwich or robust estimator

$$
V_G = \left(\sum_{i=1}^{N} D_i^{\top} V_i^{-1} D_i \right)^{-1} \left\{ \sum_{i=1}^{N} D_i^{\top} V_i^{-1} \mathrm{Cov}(Y_i) V_i^{-1} D_i \right\} \left(\sum_{i=1}^{N} D_i^{\top} V_i^{-1} D_i \right)^{-1}, \quad (4.34)
$$

where β, α, and ϕ are replaced by their estimators.

Let $\hat{\epsilon}_i = Y_i - \hat{\mu}_i$. Liang and Zeger (1986) proposed to estimate $\mathrm{Cov}(Y_i)$ in V_G by $\hat{\epsilon}_i \hat{\epsilon}_i^{\top}$, and then we have

$$
V_{LZ} = \left(\sum_{i=1}^{N} D_i^{\top} V_i^{-1} D_i \right)^{-1} \left\{ \sum_{i=1}^{N} D_i^{\top} V_i^{-1} \hat{\epsilon}_i \hat{\epsilon}_i^{\top} V_i^{-1} D_i \right\} \left(\sum_{i=1}^{N} D_i^{\top} V_i^{-1} D_i \right)^{-1}. \quad (4.35)
$$

If the covariance matrix is correctly specified, then $\mathrm{Cov}(Y_i) = V_i$. The covariance matrix estimator of $\hat{\beta}$ becomes the so-called model based covariance estimator (or naive covariance estimator),

$$
V_{MB} = \left(\sum_{i=1}^{N} D_i^{\top} V_i^{-1} D_i \right)^{-1}.
$$

Since $\hat{\epsilon}_i \hat{\epsilon}_i^{\top}$ is based on the data from only one subject, it is neither consistent nor efficient, and hence it is not an optimal estimator of $\mathrm{Cov}(Y_i)$. Although $\hat{\epsilon}_i \hat{\epsilon}_i^{T}$ is a poor estimator of $\mathrm{Cov}(Y_i)$, V_{LZ} is a consistent estimator of V_G.

An improved version of V_R estimator under certain restrictions on the data structure is given by Pan (2001). The same asymptotic normality for the β estimator in our case can be established as in Crowder (2001).

Assume that (a) the marginal variance of $\mathrm{var}(y_{ik})$ is modeled correctly, and (b) there is a common correlation structure R_0 across all subjects, then R_0 can be estimated by

$$
\hat{R}_C = \frac{1}{\phi N} \sum_{i=1}^{N} A_i^{-1/2} \hat{\epsilon}_i \hat{\epsilon}_i^{\top} A_i^{-1/2},
$$

where β, α, and ϕ are replaced by their estimators. Under the two assumptions which are often reasonable, Pan (2001) proposed estimating $\text{Cov}(Y_i)$ by

$$W_i = \phi A_i^{1/2} \hat{R}_C A_i^{1/2} = A_i^{1/2} \left(\frac{1}{N} \sum_{j=1}^{N} A_j^{-1/2} \hat{\epsilon}_j \hat{\epsilon}_j^{\top} A_j^{-1/2} \right) A_i^{1/2}.$$

Replacing $\hat{\epsilon}_i \hat{\epsilon}_i^{\top}$ in (4.35) by W_i, a new covariance matrix estimator V_P can be obtained (Pan, 2001) and is given by

$$V_P = \left(\sum_{i=1}^{N} D_i^{\top} V_i^{-1} D_i \right)^{-1} \left\{ \sum_{i=1}^{N} D_i^{\top} V_i^{-1} W_i V_i^{-1} D_i \right\} \left(\sum_{i=1}^{N} D_i^{\top} V_i^{-1} D_i \right)^{-1}. \tag{4.36}$$

Let

$$M_{LZ} = \sum_{i=1}^{N} D_i^{\top} V_i^{-1} \hat{\epsilon}_i \hat{\epsilon}_i^{\top} V_i^{-1} D_i \quad \text{and} \quad M_P = \sum_{i=1}^{N} D_i^{\top} V_i^{-1} W_i^{\top} V_i^{-1} D_i.$$

For any matrix B, define the operator $vec(B)$ as that of stacking the columns of B together to obtain a vector. Then under mild regularity conditions, $\text{cov}\{vec(M_{LZ})\} - \text{cov}\{vec(M_P)\}$ is nonnegative definite with probability 1 as $N \to +\infty$.

When the number of subjects is small (N is small), V_{LZ} would be expected underestimate $\text{var}(\hat{\beta})$ (Mancl and DeRoune, 2001). Therefore, Mancl and DeRoune (2001) proposed an alternative robust covariance estimator for $\hat{\beta}$ to reduce the bias of the residual estimator, $\hat{\epsilon}_i \hat{\epsilon}_i^{\top}$. The first-order Taylor series expansion of the residual vector $\hat{\epsilon}_i$ about β is given by

$$\hat{\epsilon}_i = \epsilon_i(\hat{\beta}) = \epsilon_i + \frac{\partial \epsilon_i}{\partial \beta^{\top}} (\hat{\beta} - \beta),$$
$$= \epsilon_i - D_i(\hat{\beta} - \beta), \tag{4.37}$$

where $\epsilon_i = \epsilon_i(\beta) = Y_i - \mu_i$ and $i = 1, \ldots, N$. Therefore,

$$\mathbf{E}(\hat{\epsilon}_i \hat{\epsilon}_i^{\top}) \approx \mathbf{E}(\epsilon_i \epsilon_i^{\top}) - \mathbf{E}[\epsilon_i(\hat{\beta} - \beta)^{\top} D_i^{\top}]$$
$$- \mathbf{E}[D_i(\hat{\beta} - \beta)\epsilon_i^{\top}] + \mathbf{E}[D_i(\hat{\beta} - \beta)(\hat{\beta} - \beta)^{\top} D_i^{\top}],$$

in which $\hat{\beta} - \beta$ can be approximated by the first order Taylor series expansion of (4.18), that is,

$$\hat{\beta} - \beta \approx \left(\sum_{i=1}^{N} D_i^{\top} V_i^{-1} D_i \right)^{-1} \sum_{i=1}^{N} D_i^{\top} V_i^{-1} \epsilon_i.$$

Hence,

$$\mathbf{E}(\hat{\epsilon}_i \hat{\epsilon}_i^{\top}) \approx \text{Cov}(Y_i) - \text{Cov}(Y_i) H_{ii}^{\top} - H_{ii} \text{Cov}(Y_i) + \sum_{j=1}^{N} H_{ij} \text{Cov}(Y_i) H_{ij}^{\top},$$

$$= (I - H_{ii}) \text{Cov}(Y_i)(I - H_{ii})^{\top} + \sum_{j \neq i} H_{ij} \text{Cov}(Y_i) H_{ij}^{\top} \tag{4.38}$$

where $H_{ij} = D_i\left(\sum_{k=1}^{N} D_k^\top V_k^{-1} D_k\right)^{-1} D_j^\top V_j^{-1}$. Since the elements of H_{ij} are between zero and one, and usually close to zero, it is reasonable to assume that the summation makes only a small contribution to the bias (Mancl and DeRoune, 2001). Therefore, $\mathbf{E}(\hat{\epsilon}_i \hat{\epsilon}_i^\top)$ can be approximated by

$$\mathbf{E}(\hat{\epsilon}_i \hat{\epsilon}_i^\top) \approx (I - H_{ii})\mathrm{Cov}(Y_i)(I - H_{ii})^\top.$$

The covariance matrix $\mathrm{Cov}(Y_i)$ can be estimated by

$$\widehat{\mathrm{Cov}(Y_i)} = (I - H_{ii})^{-1} \hat{\epsilon}_i \hat{\epsilon}_i^\top (I - H_{ii}^\top)^{-1}.$$

Mancl and DeRoune (2001) proposed the bias-corrected covariance estimator of $\hat{\beta}$:

$$V_{MD} = V_{MB}\left\{\sum_{i=1}^{N} D_i^\top V_i^{-1}(I - H_{ii})^{-1}\hat{\epsilon}_i \hat{\epsilon}_i^\top (I - H_{ii}^\top)^{-1} V_i^{-1} D_i\right\}^{-1} V_{MB}.$$

If the covariance matrix V_i is correctly specified, then (4.38) can be simplified to

$$\mathbf{E}(\hat{\epsilon}_i \hat{\epsilon}_i^\top) \approx \mathrm{Cov}(Y_i) - D_i V_{MD} D_i^\top. \tag{4.39}$$

Because $D_i V_{MD} D_i^\top$ is positive definite, the Sandwich estimate V_{LZ} appears to biased downward. Kauerman and Carroll (2001) proposed using $(I - H_{ii})^{-1/2}\hat{\epsilon}_i$ to replace $\hat{\epsilon}_i$ in V_{LZ} and gave a bias-reduced sandwich estimate

$$V_{KC} = V_{MB}\left\{\sum_{i=1}^{N} D_i^\top V_i^{-1}(I - H_{ii})^{-1/2}\hat{\epsilon}_i \hat{\epsilon}_i^\top (I - H_{ii}^\top)^{-1/2} V_i^{-1} D_i\right\}^{-1} V_{MB}.$$

In practice, it seems to be a plausible strategy to work with V_{MD} and V_{KC} instead of V_{LZ}, even if V_i is a working covariance and the true variance structure is unknown.

The basic idea of the three modified covariance matrix estimators is to seek a good estimator of the covariance matrix $\mathrm{Cov}(Y_i)$. The sandwich estimator is proposed in the case of misspecified models. There has been some great theory established by Eicker (1963); Huber (1967); and White (1982).

The statistical properties of these improved estimators of the covariance matrix mentioned above are not discussed here. More details can be found in references Pan (2001), Kauerman and Carroll (2001), and Mancl and DeRoune (2001).

4.4.5 Example: Epileptic Data

Thall and Vail (1990) analyzed the data from a clinical trial of 59 epileptics using various variance functions and correlation structures. Patients were randomized to receive either the anti-epileptic drug progabide or a placebo. For each patient, the number of epileptic seizures was recorded during a baseline period of eight weeks. During the treatment, the number of seizures was then recorded in four consecutive two-week intervals. The data also includes age of patients, treatment indicator with 1 for progabide and 0 for placebo. The medical interest is whether or not the progabide reduces the rate of epileptic seizures.

The data shows high degree of extra-Poisson variation and within subject correlation as demonstrated in Thall and Vail (1990) and Diggle et al. (2002). Interesting work has been done in Thall and Vail (1990). We will apply the proposed approach assuming different variance functions including the power function and compare with results from other competing models.

We consider five covariates including intercept, treatment, baseline seizure rate, age of subject, and the interaction between treatment and baseline seizure rate. The mean vector for the ith subject is $\mu_i = \exp(x_i^T \beta)$, where x_i is the design matrix for the covariates. They transformed baseline as the logarithm of $1/4$ of the 8-week pre-randomization seizure count and log-transformed age. The treatment variable is a binary indicator for the progabide group.

The overdispersion value ϕ for the two treatment groups estimated by the ratio between sample variance and mean are far away from 1, demonstrating the high degree of extra-Poisson variation. The sample mean-variance plot also exhibits some nonlinear trend. We, therefore, assume two variance functions for the data: one is quadratic function $\sigma_{ij}^2 = \gamma_1 \mu_{ij} + \gamma_2 \mu_{ij}^2$, which was introduced by Bartlett (1936) and Morton (1987); the other is power function $\sigma_{ij}^2 = \phi \mu_{ij}^\gamma$ with γ being estimated from the data. Paul and Plackett (1978) suggested both of these functions for overdispersed Poisson data. Two different working correlation structures, AR(1) and the exchangeable, are used to take account of within subject correlations. The proposed approach is applied to perform the estimation of variance and regression parameters. The correlation parameters are estimated by the moment methods. The final estimates of the parameters are obtained after a few iterations among regression, variance and correlation parameters. The asymptotic covariance matrix of β is estimated by the sandwich estimator.

Table 4.2 shows the summary statistics for the two-week seizure counts. Overdispersion to the Poisson model is evident because the ratio of variance to mean is much larger than 1. We will further discuss this example in Chapter 5.

Once the Quasi-poisson (Poisson with overdispersion) model is fitted, we obtained the fitted values $\hat{\mu}_{ik}$, and residuals r_{ik}. The variance parameters are then simply obtained by minimizing the $\sum_{i,k}\{r_{ik}^2 - \sigma_{ik}^2\}^2/\hat{\mu}_{ik}^2$ for the nominated variance function (Power or Bartlett). The resultant variance function can be used as the weight function weights=μ_{ik}/σ_{ik}^2 and family = poisson to make use of the existing geese function (*geepack*). Further iterations can also be employed in updating the variance parameters and the weights as an exercise.

Table 4.3 shows the results for different variance functions. The overdispersion parameter ϕ values obtained from different correlation structures are very similar for a given variance function.

It should be noted that in this example all covariates are at cluster level, i.e., they do not change within each subject. In these cases, the constant correlation assumption (exchangeable structure) will lead to the same estimates as the independence models. Even the standard errors are the same. This is because they are obtained by the robust sandwich approach (instead of the naive one). It is clear that choice of variance

Table 4.2 *Mean and variance for the two-week seizure count within each group. The ratio of variance to mean shows the extent of Poisson overdispersion ($\hat{\phi} = s^2/\bar{Y}$).*

Visit	Placebo ($M_1 = 28$)			Progabide ($M_2 = 31$)		
	$\bar{Y}$	s^2	$\hat{\phi}$	$\bar{Y}$	s^2	$\hat{\phi}$
1	9.36	102.76	10.98	8.58	332.72	38.78
2	8.29	66.66	8.04	8.42	140.65	16.70
3	8.79	215.29	23.09	8.13	193.05	23.75
4	7.96	58.18	7.31	6.71	126.88	18.91

function have a great impact on the estimates and especially on the standard error. When the Bartlett and the power variance are used, the estimates have much smaller standard errors. Estimates and their standard errors from the Bartlett variance model differ little for different working correlation structures. However, across the nine covariance models, the standard errors ranged from 0.174 to 0.211 for the interaction term, and 0.415 to 0.464 for the treatment. This demonstrates that attention in modeling the variance function is necessary instead of just using the default variance functions to achieve high estimation efficiency. In this example, further goodness-of-fit development is needed for assessing different variance and correlation models. Model selection criteria will be discussed in the coming Chapter.

4.4.6   Infeasibility

Crowder (1995) found that under misspecification of the correlation structure, the designated working correlation matrix may not even converge to a correlation matrix. This generates interest in searching for better estimates of correlation parameters or alternatives to the working matrix method (Chaganty, 1997; Qu, Lindsay & Li, 2000). Crowder (1995) redefines GEE as seeking simultaneous solutions of estimating equations $\mathbf{U}_\beta(\theta) = 0$ and $\mathbf{U}_\alpha(\theta) = 0$. Crowder shows that "lack of a parametric status of α" leads to indefiniteness of underlying stochastic structure for the outcome vectors, so that the $\sqrt{K}$-consistency of $\hat{\alpha}$ required to obtain favorable properties of solutions to $\mathbf{U}_\beta(\theta) = 0$ may be meaningless. Crowder illustrates this indefiniteness by attempting to solve a GEE with working structure $\mathbf{R}(\alpha)$ when the data arise from a model with true cluster correlation structure $\tilde{\mathbf{R}}$ that is very different from $(\mathbf{R})$. Specifically, when $(\tilde{R})_{jk} = \alpha$ (exchangeable true correlation) but $(\mathbf{R})_{jk} = \alpha^{|j-k|}$ (autoregressive working correlation), then for $n_i = 3$ and $-1/2 \leq \alpha < -1/3$, an obvious moment-based formulation of $\mathbf{U}_\alpha(\theta)$ has no real-valued root between -1 and 1. This destroys prospects for a general theory of existence and consistency of simultaneous solutions to the estimating functions. Crowder argues that this indefiniteness also infects the GEE2 procedures of Prentice and Zhao (1991).

   In a seminal study, Crowder (1995) indicates that while robustness to misspecification of $cov(Y_i)$ is a key attraction of the GEE approach, study of performance under misspecification requires additional formalization. The ad hoc estimators of Liang and Zeger are, therefore, re-expressed as supplemental estimating equations whose solutions $\tilde{\alpha}$ can be characterized under misspecified structure for $cov(Y_i)$. Even if α

Table 4.3 *Parameters estimates from models with different variance functions and correlation structures for the epileptic data.*

	log(baseline)	log(age)	trt	intact
Poisson Variance: $\sigma_{ij}^2 = \phi\mu_{ij}$.				
Independence working model,$\hat{\phi} = \exp(1.486)$				
β	0.950	0.897	-1.341	0.562
Stderr	0.097	0.275	0.426	0.174
AR(1) working model, $\hat{\phi} = \exp(1.509), \hat{\alpha} = 0.495$				
β	0.941	0.994	-1.502	0.626
Stderr	0.091	0.272	0.415	0.168
Exchangeable working model, $\hat{\phi} = \exp(1.486), \hat{\alpha} = 0.347$				
β	0.950	0.897	-1.341	0.562
Stderr	0.097	0.275	0.426	0.174
Bartlett variance: $\sigma_{ij}^2 = 3.424*\mu_{ij} + 0.120*\mu_{ij}^2$.				
Independence working model, $\hat{\phi} = \exp(1.402)$				
β	0.943	0.762	-1.074	0.434
Stderr	0.110	0.266	0.464	0.211
AR(1) working model, $\hat{\phi} = \exp(1.422), \hat{\alpha} = 0.477$				
β	0.933	0.850	-1.219	0.492
Stderr	0.102	0.264	0.458	0.208
Exchangeable working model, $\hat{\phi} = \exp(1.33), \hat{\alpha} = 0.294$				
β	0.943	0.762	-1.074	0.434
Stderr	0.110	0.266	0.464	0.211
Power function variance: $\sigma_{ij}^2 = \phi\mu_{ij}^{1.329}$.				
Independence working model, $\hat{\phi} = \exp(1.486)$				
β	0.924	0.719	-1.091	0.449
Stderr	0.109	0.267	0.431	0.198
AR(1) working model, $\hat{\phi} = \exp(1.348), \hat{\alpha} = 0.438$				
β	0.915	0.796	-1.210	0.496
Stderr	0.102	0.265	0.429	0.198
Exchangeable working model, $\hat{\phi} = \exp(1.33), \hat{\alpha} = 0.294$				
β	0.924	0.719	-1.091	0.449
Stderr	0.109	0.267	0.431	0.198

is regarded strictly as a nuisance parameter, the effects of misspecification of $\text{cov}(Y_i)$ and the subsequent misweighting of the observations may propagate to properties of $\hat{\beta}_G$. Specifically, Crowder shows that if $\text{cov}(Y_i)$ has the AR(1) form, but the working model is chosen to have the exchangeable (compound symmetric) form then (for a balanced design) the limit of the moment estimator $\tilde{\alpha}$ depends explicitly on the cluster size. Additionally, if $\text{cov}(Y_i)$ has the exchangeable form but the working model is of AR(1) structure, then for $n_i \equiv 3$, if $-1/2 \leq \alpha_{exch} \leq -1/3$ the estimating equation associated with the AR(1) moment estimator has no real solution, and neither does the GEE. Crowder concludes that "there can be no general asymptotic theory supporting existence and consistency of $(\hat{\beta}, \hat{\alpha})$" ... "$\alpha$ has no underlying parametric identity independent of the particular estimating equation chosen: α does not exist in any fundamental sense." Crowder concludes his discussion with the positive recommendation to use only estimating equations that have guaranteed solution, or estimate α by minimizing a well-behaved objective function. He also notes that in practice "statistical judgement would normally be employed in an attempt to avoid such hidden pitfalls" as infeasible estimators obtained through misspecification. In a follow-up to Crowder's paper, Sutradahar and Das (1999) argued that solutions to GEEs using the (typically frankly misspecified) working independence model can often be more efficient than those obtained under misspecified nonindependence working models. To support their argument they investigate a model for binary outcomes

$$\text{logit } EY_{it} = \beta_0 + \beta_1 t$$

with balanced design $n_i \equiv n$. Let $\hat{\beta}_{struct}$ denote the solution to the GEE with working correlation structure $struct \in \{indep, exch, AR(1)\}$, and let eff $\hat{\beta}_{struct}$ denote V_{true}/V_{struct}, where V_{struct} is the asymptotic variance of a component of the solution to the GEE with working correlation structure $struct$. When true $\text{cov}(Y_i)$ is of AR(1) form but the working model is exchangeable, they show that eff $(\hat{\beta}_{indep}) \approx$ eff $(\hat{\beta}_{exch})$. When the true $\text{cov}(Y_i)$ is of exchangeable form but the working model is AR(1), they show that eff $(\hat{\beta}_{indep}) \gg$ eff $(\hat{\beta}_{AR(1)})$. They conclude that the "use of working GEE estimator $\hat{\beta}_G$ may be counterproductive", and that their "results contradict the recommendations made in LZ that one should use $\hat{\beta}_G$ for higher efficiency relative to $\hat{\beta}_{indep}$."

Investigations of infeasibility and efficiency loss under misspecification undertaken thus far do not acknowledge the additional complications of unbalanced designs and within/between-cluster covariate dispersion. The latter phenomenon has been studied by Fitzmaurice (1995), who showed that the relative efficiency of solutions obtained under working independence depends critically on the within-cluster correlation of covariates. Furthermore, no study of tools for data-based selection of working correlation models has been undertaken to date. A reliable correlation structure selection method would greatly diminish the reservations concerning feasibility and efficiency loss reviewed here.

To best illustrate the problem raised by Crowder (1995), let us revisit his example in which the true correlation structure is exchangeable with $n_i = 3$ and the AR(1) working correlation matrix is used. To be more specific, we use

$$R_i = \begin{pmatrix} 1 & \alpha & \alpha^2 \\ \alpha & 1 & \alpha \\ \alpha^2 & \alpha & 1 \end{pmatrix}$$

while the true correlation matrix is

$$\tilde{R}_i = \begin{pmatrix} 1 & \rho & \rho \\ \rho & 1 & \rho \\ \rho & \rho & 1 \end{pmatrix}.$$

If we use the simple moment estimator as given by (4.22) or the Cholesky estimator 4.32), their limit converges to ρ, the true exchangeable correlation parameter
value, i.e., asymptotically, we will use an incorrect AR(1) correlation matrix although
R_i will never be correct no matter what α is used (even if $\alpha = \rho$). In these cases, there
is no problem. However, if we use a moment estimator that is based on all the residual
products, the estimating equation is

$$q_\alpha = \sum_{i=1}^{n} (\hat{\epsilon}_{i1}\hat{\epsilon}_{i2} + \hat{\epsilon}_{i2}\hat{\epsilon}_{i3} + \hat{\epsilon}_{i1}\hat{\epsilon}_{i3}) - n\phi(2\alpha + \alpha^2) = 0,$$

where $\hat{\epsilon}_{ij} = (y_{ij} - \mu_{ij})/(A_i)_{jj}$, and $(A_i)_{jj}$ is the j-the element of matrix A_i.
   A reasonable estimator of ϕ is

$$\hat{\phi} = \frac{1}{n} \sum_{i=1}^{n} (\hat{\epsilon}_{i1}^2 + \hat{\epsilon}_{i2}^2 + \hat{\epsilon}_{i3}^2)/3.$$

For convenience, let

$$\hat{\rho} = \frac{\sum_{i=1}^{n} (\hat{\epsilon}_{i1}\hat{\epsilon}_{i2} + \hat{\epsilon}_{i2}\hat{\epsilon}_{i3} + \hat{\epsilon}_{i1}\hat{\epsilon}_{i3})}{3n\hat{\phi}} = \frac{\sum_{i=1}^{n} (\hat{\epsilon}_{i1}\hat{\epsilon}_{i2} + \hat{\epsilon}_{i2}\hat{\epsilon}_{i3} + \hat{\epsilon}_{i1}\hat{\epsilon}_{i3})}{\sum_{i=1}^{n} (\hat{\epsilon}_{i1}^2 + \hat{\epsilon}_{i2}^2 + \hat{\epsilon}_{i3}^2)}.$$

The solution of $q_\alpha = 0$ is given by

$$\hat{\alpha}_1 = \begin{cases} \sqrt{1+3\hat{\rho}} - 1 & \text{if } \hat{\rho} > -1/3 \\ \text{undefined} & \text{if } \hat{\rho} \leq -1/3 \end{cases}$$

Clearly, $\hat{\rho}$ is a random variable taking values between -1 and 1, and it is quite possible
that $\text{pr}(\hat{\rho} < -1/3) > 0$. In fact, if the true correlation structure is exchangeable with a
parameter $\rho < -1/3$, we have $\text{pr}(\hat{\rho} < -1/3) \to 1$ as $n \to +\infty$, indicating that $\hat{\alpha}$ would
be undefined with probability 1.
   Crowder (1995) further argued that the limiting value of $\hat{\alpha}$ may not exist and
hence the asymptotic properties for the GEE estimators breakdown. The main con-
cern here is we cannot guarantee the condition $n^{1/2}(\hat{\alpha} - \alpha_0) = O_p(1)$ holds, which
is required for the existence of $\hat{\alpha}$ and its limiting value α_0. The implication of this
concern is that it is important to ensure the $\hat{\alpha}$ has the asymptotic normality. Care-
fully chosen supplementary estimating functions will not have this issue, and choice

of a "working" correlation should be as close to the truth as possible. In the above example, even if the true model is AR(1) but with $\rho \leq -1/3$, $\hat{\alpha}$ can still be problematic. This indicates the moment estimator using all pairwise product is statistically problematic for the AR(1) model. As we mentioned, the simple lag-1 estimator

$$\hat{\gamma}_2 = \frac{2\sum_{i=1}^{n}(\hat{\epsilon}_{i1}\hat{\epsilon}_{i2} + \hat{\epsilon}_{i2}\hat{\epsilon}_{i3})}{\sum_{i=1}^{n}(\hat{\epsilon}_{i1}^2 + 2\hat{\epsilon}_{i2}^2 + \hat{\epsilon}_{i3}^2)},$$

does not have the aforementioned infeasibility issue either regardless the true model is exchangeable or AR(1).

Of course, in cases when no sensible correlation matrix is produced by a nominated "working" model, we should embark on the other working models (including the independence model).

In the above example, when we have $\hat{\rho} \leq -1/3$, one can consider a different estimator for α,

$$\hat{\alpha}_2 = \begin{cases} \sqrt{1+3\hat{\rho}} - 1 & \text{if } \hat{\rho} \geq -1/3 \\ \hat{\gamma}_1 & \text{if } \hat{\rho} < -1/3 \end{cases}, \tag{4.40}$$

in which $\hat{\gamma}_1 = \sum_{i=1}^{n}(\hat{\epsilon}_{i1}\hat{\epsilon}_{i2} + \hat{\epsilon}_{i2}\hat{\epsilon}_{i3})/(2n\hat{\phi})$. The infeasibility issue also disappears when the working matrix is AR(1) with a parameter $\hat{\alpha}_2$.

Alternatively, if we choose a different working structure, say, equicorrelated with a parameter γ which can be estimated by products of all pairs,

$$\hat{\gamma}_2 = \sum_{i=1}^{n}(\hat{\epsilon}_{i1}\hat{\epsilon}_{i2} + \hat{\epsilon}_{i2}\hat{\epsilon}_{i3} + \hat{\epsilon}_{i1}\hat{\epsilon}_{i3})/(3n\hat{\phi}).$$

In this case, $\hat{\alpha}$ also takes the same form as (4.40) replacing $\hat{\gamma}_1$ by $\hat{\gamma}_2$, but the corresponding working matrix involves two different structures: AR(1) with the parameter $\sqrt{1+3\hat{\rho}} - 1$ if $\hat{\rho} > -1/3$ and equicorrelated matrix with the parameter $\hat{\gamma}_2$. In practice, one can also consider independence ($\hat{\alpha} = 0$) when the nominated parametric correlation matrix is not positive definite.

The conclusion is that this infeasibility is a theoretic concern, and it does not happen in practice, not even rarely.

4.5   Quadratic Inference Function

As one can see supplying a sensible "working" correlation matrix and the efficient estimating functions for the corresponding α parameters can be challenges. This motivates us to consider other approaches such as the random effects models, or Transitional GEE approach in which the "working" covariance matrix is always diagonal.

Alternatively, one can also consider a series of "working" models, and establish model selection criteria to help us identify the best models. We will provide more details on this in the next chapter.

Apart from these considerations, one can also consider optimally combine a number of nominated working models to bypass the need to specify one particular model each time. This leads to the question, can we generalize the unstructured working correlation models so that the estimating functions can automatically combine all possible structures?

Suppose we have set of candidate correlation models, the K sets of estimating functions for estimating β,

$$U(\beta) = \sum_{i=1}^{N} \begin{pmatrix} D_i^\top A_i^{-1/2} R_i^{(1)} A_i^{-1/2} (Y_i - \mu_i) \\ D_i^\top A_i^{-1/2} R_i^{(2)} A_i^{-1/2} (Y_i - \mu_i) \\ \dots \\ D_i^\top A_i^{-1/2} R_i^{(K)} A_i^{-1/2} (Y_i - \mu_i) \end{pmatrix} = \sum_{i=1}^{N} U_i(\beta),$$

in which U_i has Kp components denoting the contributions from subject i. Note that $R_i^{(k)}$ is symmetric with non-negative eigenvalues. Applying the generalized method of moment, we can obtain the optimal p-estimating functions for β of dimension p. The generalized moment method can be applied to combine all these together for estimating β (Hansen, 1982). The Quadratic inference function for β proposed by Qu et al. (2000) is to minimize

$$U^\top(\beta) \widehat{\mathrm{cov}(U)} U(\beta),$$

where $\widehat{\mathrm{cov}(U)}$ is the emprical estimate of the covariance matrix of $U(\beta)$, $\sum_{i=1}^{N} U_i(\beta) U_i^\top(\beta)$. A number of interesting properties and extensions were established by Qu et al. (2000); Qu and Song (2004); Qu et al. (2008); Wang and Qu (2009). An excellent review is given by Song et al. (2009). In particular, an interesting extension is made for generalized mixed-effects models by Wang et al. (2012b) with penalized random effects in the estimating functions U_i.

There is a wide class available for the correlation structure for R_i available in argument *correlation = ,* and also for the within-group heteroscedasticity structure using *weights=.*

Chapter 5

Model Selection

5.1 Introduction

Model selection is an important issue in almost any practical data analysis. For longitudinal data analysis, model selection commonly includes covariate selection in regression and correlation structure selection in the GEE discussed in Chapter 4. Covariate selection in regression means: given a large group of covariates (including some higher-order terms or interactive effects), one needs to select a subset to be included in the regression model. Correlation structure selection means: given a pool of working correlation structure candidates, select one that is closer to the truth, and thus results in more efficient parameter estimates. Such as the study of the National Longitudinal Survey of Labor Market Experience introduced in Chapter 1, covariates including age, grade, and south, were recorded on the participants over the years. Even though the number of variables is not large in this example, when various interaction effects are included, the total number of covariates in the statistical model can be considerably large. However, only a subset of them is relevant to the response variable. Inclusion of redundant variables may hinder accuracy and efficiency for both estimation and inference. Furthermore, a correlation structure needs to specify when using the GEE to estimate parameters. Appropriate specification of correlation structures in longitudinal data analysis can improve estimation efficiency and lead to more reliable statistical inferences. Thus, it is important to use statistical methodology to select important covariates and an appropriate correlation structure. There is a lot of model-selection literature in statistics (e.g., Miller, 1990, Jiang et al. (2015) and references therein) but mainly for covariate selection in the classic linear regression with independent data. Traditional model selection criteria such as the Akaike information criterion (AIC), and the Bayesian information criterion (BIC) may not be useful for correlation structure selection because the joint density function of response variables is usually unknown in longitudinal data.

Suppose that a longitudinal dataset composed of outcome variables $Y_i = (y_{i1}, \ldots, y_{in_i})^{\mathrm{T}}$, and corresponding covariate variables $X_i = (X_{i1}, \ldots, X_{in_i})^{\mathrm{T}}$ in which X_{ij} be a $p \times 1$ vector, $i = 1, \ldots, N$. For the sake of simplicity, we assume that $n_i = n$ for all i and the total number of observations is $M = Nn$. Assumed that

$$(i) \quad \mu_{ij} = g(X_{ij}^{\mathrm{T}} \beta), \tag{5.1}$$

DOI: 10.1201/9781315153636-5

where $g(\cdot)$ is a specified, function, and β is an unknown parameter vector, and that

$$(ii) \quad \sigma_{ij}^2 = \phi v(\mu_{ij}), \tag{5.2}$$

where $v(\cdot)$ is a given function of μ_{ij}, and ϕ is a scale parameter. Assume that the co-variance matrix of Y_i is $V_i = \phi \Sigma_i$ in which $\Sigma_i = A_i^{1/2} R_i(\alpha) A_i^{1/2}$, where $A_i = \text{diag}(v(\mu_{ij}))$ and $R_i(\alpha)$ is the true correlation matrix of Y_i with an unknown $r \times 1$ parameter vector α.

Based on the assumptions (5.1) and (5.2), the GEE is

$$U(\beta) = \sum_{i=1}^{N} D_i^T W_i^{-1} (Y_i - \mu_i) = 0, \tag{5.3}$$

where $D_i = \partial \mu_i / \partial \beta$, and $W_i = \phi A_i^{1/2} R_w A_i^{1/2}$, in which R_w is a working correlation matrix of Y_i. Define $\hat{\beta}_R$ is the resulting estimator from the equation (5.3), then the robust variance estimator of $\hat{\beta}_R$ is

$$V_R = \Omega_R^{-1} \left(\sum_{i=1}^{N} D_i^T W_i^{-1} \text{var}(Y_i) W_i^{-1} D_i \right) \Omega_R^{-1},$$

where

$$\Omega_R = \sum_{i=1}^{N} D_i^T W_i^{-1} D_i$$

is the model-based estimator of $\text{var}(\hat{\beta}_R)$. Specifically, if we specify R_w as the identify matrix, then Ω_R^{-1} reduces to

$$\Omega_I = \sum_{i=1}^{N} D_i^T A_i^{-1} D_i / \phi.$$

A consistent estimator of V_R is

$$\hat{V}_R = \Omega_R^{-1} \left(\sum_{i=1}^{N} D_i^T W_i^{-1} S_i S_i^T W_i^{-1} D_i \right) \Omega_R^{-1} \Bigg|_{\beta = \hat{\beta}_R, \alpha = \hat{\alpha}_R, \phi = \hat{\phi}_R}$$

in which $\hat{\alpha}_R$ and $\hat{\phi}_R$ are consistent estimators of α and ϕ. In the following sections, we mainly introduce several criteria for selecting covariates in (5.1) and choosing an appropriate correlation structure for R_w in (5.3) for longitudinal data analysis.

5.2   Selecting Covariates

5.2.1   Quasi-likelihood Criterion

The AIC is based on the likelihood function and asymptotic properties of the maximum likelihood estimator. Since no distribution is assumed in the GEE, there is no likelihood is defined, and thus the AIC cannot be directly used. Pan (2001) proposed a criterion that is an extension of the AIC based on quasi-likelihood, named the QIC, to choose the appropriate mean model or the working correlation structure.

Based on the model specification $\mu_{ij} = \mathbf{E}(y_{ij}|x_{ij})$ and $\sigma_{ij}^2 = \phi v(\mu_{ij})$, the quasi-log-likelihood function of y_{ij} is

$$Q(y_{ij};\mu_{ij},\phi) = \int_{y_{ij}}^{\mu_{ij}} \frac{y_{ij}-t}{\phi v(t)} dt.$$

If y_{ij} is a continuous response with specified $v(\mu_{ij}) = 1$, then $Q(y_{ij},\mu_{ij},\phi) = -(y_{ij} - \mu_{ij})^2/(2\phi)$. If y_{ij} is a binary response and with specified $v(\mu_{ij}) = \mu_{ij}(1 - \mu_{ij})$, then $Q(y_{ij},\mu_{ij},\phi) = y_{ij}\log\{\mu_{ij}/(1 - \mu_{ij})\} + \log(1 - \mu_{ij})$. If y_{ij} is a count response with specified $v(\mu_{ij}) = \mu_{ij}$, then $Q(y_{ij},\mu_{ij},\phi) = (y_{ij}\log\mu_{ij} - \mu_{ij})/\phi$ (McCullagh and Nelder, 1989). The dispersion parameter $\phi = 1$ for a binary response. For other response types, ϕ is unknown, and $\phi > 1$ is extremely useful in modeling over-dispersion that commonly occurs in practice.

If we assume that the working correlation matrix R_w is the identify matrix I, then the quasi-log-likelihood is

$$Q(\beta,\phi;I) = \sum_{i=1}^{N} \sum_{j=1}^{n} Q(y_{ij};\mu_{ij},\phi).$$

Therefore, for a given working correlation matrix R, the QIC can be expressed as

$$\mathrm{QIC(R)} = -2Q(\hat{\beta}_R,\hat{\phi};I) + 2\mathrm{tr}(\hat{Q}_I\hat{V}_R), \tag{5.4}$$

where the first term is the estimated quasi-log-likelihood with $\beta = \hat{\beta}_R$. The second term is the trace of the product of $\hat{Q}_I = \phi^{-1}\sum_{i=1}^{N} D_i^T A_i^{-1} D_i|_{\beta=\hat{\beta}_R,\phi=\hat{\phi}_R}$ and $\hat{V}_R$ given in Section 1. Here $\hat{Q}_I$ and $\hat{V}_R$ can be directly available from the model fitting results in many statistical softwares, such as SAS, S-Plus, and R. Besides selecting covariates in generalized linear models, the QIC can also be applied to select a working correlation structure in GEE: one needs to calculate the QIC for various candidate working correlation structures and then pick the one with the smallest QIC.

In practice, since ϕ is unknown, we estimate it using $\hat{\phi} = (M-p)^{-1}\sum_{i=1}^{N}\sum_{k=1}^{n}(y_{ik} - \hat{\mu}_{ik})^2/v(\hat{\mu}_{ik})$, in which $\hat{\mu}_{ik}$ is estimated based on the regression model including all covariates. When all modeling specifications in GEE are correct, $\hat{Q}_I^{-1}$ and $\hat{V}_R$ are asymptotically equivalent and $\mathrm{tr}(\hat{Q}_I\hat{V}_R) \approx p$. Then the QIC reduces to the AIC. In GEE with longitudinal data, one may take $\mathrm{QIC}_u(R) = -2Q(\hat{\beta}_R,I) + 2p$ as an approximation to the QIC(R), and thus QIC(R) can be potentially useful in variable selection. However, $\mathrm{QIC}_u(R)$ cannot be applied to select the working correlation matrix R. That is because the value of $Q(\hat{\beta}_R,I)$ does not depend on the correlation matrix R. It is worth noting that the performance of the QICs with different correlation structures is close for selecting covariates, but the QIC(I) performs the best (Pan, 2001).

Hardin and Hilbe (2012) proposed a slightly different version of QIC(R):

$$\text{QIC}_H(R) = -2Q(\hat{\beta}_R, \hat{\phi}; I) + 2\text{tr}(\hat{\Omega}_I \hat{V}_R)$$

where $Q(\hat{\beta}_R; \hat{\phi}; I)$ and $\hat{V}_R$ are the same as those in (5.4). However, $\hat{\Omega}_I$ is evaluated using $\hat{\beta}_I$ and $\hat{\phi}_I$ instead of $\hat{\beta}_R$ and $\hat{\phi}_R$.

5.2.2   Gaussian Likelihood Criterion

Assume that Y_i is a continuous variable and follows a multivariate normal distribution with mean $X_i\beta$ and covariance matrix $\phi\Sigma_i$ for $i = 1,\dots,N$, then the log-likelihood function of $Y = (Y_1^T,\dots,Y_N^T)^T$ (omitting constant terms) is

$$l(\beta,\alpha,\phi) = -\frac{1}{2}\Big[M\log\phi + \log|V| + (Y - X\beta)^T V^{-1}(Y - X\beta)/\phi\Big],$$

where V is an $M \times M$ block-diagonal matrix with $N \times N$ blocks Σ_i. Therefore, the AIC and BIC for the longitudinal data are

$$\text{AIC} = -2l(\hat{\beta},\hat{\alpha},\hat{\phi}) + 2p,$$
$$\text{BIC} = -2l(\hat{\beta},\hat{\alpha},\hat{\phi}) + p\log M,$$

where $(\hat{\beta},\hat{\alpha},\hat{\phi})$ are the maximum likelihood estimators of (β,α,ϕ). Specifically, $\hat{\beta} = (X^T\hat{V}^{-1}X)^{-1}X^T\hat{V}^{-1}Y$ and $\hat{\phi} = (Y - X\hat{\beta})^T\hat{V}^{-1}(Y - X\hat{\beta})/M$, in which $\hat{V}$ is evaluated at $\hat{\alpha}$ which is estimated by maximizing the profile log-likelihood $l_p(\alpha) = -1/2[M\log\hat{\phi}(\alpha) + \log|V(\alpha)|]$. When the sample size N is small, the corrected AIC for continuous longitudinal data is

$$\text{AICc} = -2l(\hat{\beta},\hat{\alpha},\hat{\phi}) + 2p\frac{M}{M - p - 1}.$$

When N tends to infinity with a fixed p, AICc approximates to AIC.

If Y_i is a discrete variable, Y_i does not follow a multivariate normal distribution. However, White (1961) and Crowder (1985) proposed using a normal log-likelihood to estimate β and α without assuming that Y_i is normally distributed. Hence, we can utilize the following normal log-likelihood function as a pseudo-likelihood,

$$l_G(\beta,\alpha,\phi) = -\frac{1}{2}\sum_{i=1}^{N}\Big[\log|2\pi V_i| + (Y_i - \mu_i)^T V_i^{-1}(Y_i - \mu_i)\Big],$$

where $\mu_i = (g(X_{i1}^T\beta),\dots,g(X_{in_i}^T\beta))^T$. Therefore, the AIC and BIC for the longitudinal data are

$$\text{GAIC} = -2l_G(\hat{\beta}_G,\hat{\alpha}_G,\hat{\phi}_G) + 2p,$$
$$\text{GBIC} = -2l_G(\hat{\beta}_G,\hat{\alpha}_G,\hat{\phi}_G) + p\log M,$$

where $(\hat{\beta}_G, \hat{\alpha}_G)$ are the Gaussian estimators of (β, α) introduced in **Chapter 4**, and $\hat{\phi}_G$ is obtained by $\sum_{i=1}^{N}(Y_i - \mu_i(\hat{\beta}_G))^T\hat{\Sigma}_i^{-1}(\hat{\beta}_G, \hat{\alpha}_G)(Y_i - \mu_i(\hat{\beta}_G))/M$. Note that there is a subtle difference between the traditional AIC/BIC and the GAIC/GBIC: the traditional AIC/BIC requires the likelihood function to be correct, while the GAIC/GBIC does not. Here, l_G is a working likelihood. Furthermore, the l_G in the GAIC and GBIC can be replaced by the restricted Gaussian likelihood proposed by Patterson and Thompson (1971). When the GAIC and the GBIC are used to select covariates, the simulation results indicate that the GAIC and the GBIC with the independence correlation structure perform the best.

5.3   Selecting Correlation Structure

Selecting an appropriate working correlation structure is pertinent to longitudinal data analysis using generalized estimating equations. An inappropriate correlation structure will lead to inefficient parameter estimation in the GEE framework (Hin and Wang, 2009). Besides the QIC, the GAIC, and the GBIC, we will introduce another two criteria for working correlation structure selection. Note that the goal of selecting a working correlation structure is to estimate parameters more efficiently.

5.3.1   CIC Criterion

Hin and Wang (2009) noted that the expectation of the first term in the QIC is free from the working correlation matrix R_w and the true correlation matrix R_T. Therefore, $Q(\hat{\beta}, \hat{\phi}; I, \mathcal{D})$ does not contain information about the hypothesized within correlation structure, and the random errors from $Q(\hat{\beta}, \hat{\phi}; I, \mathcal{D})$ can affect the performance of the QIC. Therefore, Hin and Wang (2009) proposed using only the second term in the QIC as a correlation information criterion (CIC) for correlation structure selection,

$$\text{CIC}(R) = \text{tr}(\hat{Q}_I \hat{V}_R).$$

Thus, $\text{QIC}(R) = -2Q(\hat{\beta}_R; I) + 2\text{CIC}(R)$. Without the effect of the random error from the first term in (5.4), the CIC could be more powerful than the QIC.

The theoretical underpinning for the biased first term in (5.4) was outlined in Wang and Hin (2010). Note that for continuous responses and ϕ estimated by $\hat{\phi}$, we have $\text{QIC}(R) = (M - p) + 2\text{CIC}(R)$, which means that the CIC is equivalent to the QIC in the working correlation structure chosen for continuous responses. If the true correlation matrix is the identify matrix I, $\hat{Q}_I^{-1}$ and $\hat{V}_R$ are asymptotically equivalent, hence $\text{CIC}(R) \approx 2p$.

5.3.2   C(R) Criterion

From a theoretical point of view, when R_w is close to the true correlation matrix R_T, the discrepancy between the covariance matrix estimator of Y_i and the specified working covariance matrix W_i is small. Therefore, according to a statistic to test the hypothesis that the covariance matrix equals a given matrix, Gosho et al. (2011) proposed a criterion to choose a working correlation structure

$$C(R) = \text{tr}\left\{\left[\left(\frac{1}{N}\sum_{i=1}^{N}S_iS_i^{\mathsf{T}}\right)\left(\frac{1}{N}\sum_{i=1}^{N}W_i\right)^{-1} - I_n\right]^2\right\}. \tag{5.5}$$

The value of the C(R) should be close to zero when R_w is accurately specified. However, the C(R) is appropriate only for the balanced data.

5.3.3 Empirical Likelihood Criteria

Suppose that $x_1,\ldots,x_N$ are independent random samples from a distribution F with a parameter vector $\theta \in \Theta \subset \mathbb{R}^p$. The empirical likelihood function is

$$L(F) = \prod_{i=1}^{N}F(x_i) = \prod_{i=1}^{N}p_i,$$

where $p_i = dF(x_i) = P(X = x_i)$. The empirical likelihood ratio is defined as (Owen, 2001)

$$\mathcal{R}(F) = L(F)/L(F_n) = \prod_{i=1}^{N}Np_i, \tag{5.6}$$

where $F_n(x) = 1/N\sum_{i=1}^{N}I(x_i \le x)$.

Assume that θ satisfies unbiased estimating equations

$$\mathbf{E}_F(g(X,\theta)) = 0, \tag{5.7}$$

where $g(\cdot) = (g_1(X,\theta),\ldots,g_r(X,\theta))^{\mathsf{T}}$ is r-dimension functions and $r \ge p$. The empirical likelihood ratio function for θ is defined as (Qin and Lawless (1994)

$$\mathcal{R}(\theta) = \sup\left\{\prod_{i=1}^{N}Np_i; 0 \le p_i \le 1, \sum_{i=1}^{N}p_i = 1, \sum_{i=1}^{N}p_ig(x_i,\theta) = 0\right\}.$$

An explicit expression value for $\mathcal{R}(\theta)$ can be derived by the Lagrange multiplier method. The maximum empirical likelihood estimator of θ is

$$\hat{\theta}_{EL} = \text{argmax}_\theta\mathcal{R}(\theta).$$

Qin and Lawless (1994) proved that $-2\log(\mathcal{R}(\theta)/\mathcal{R}(\hat{\theta})) \to \chi_p^2$, and hence confidence regions can be constructed for θ. Note that when $r = p$, it seen that $\hat{\beta}_{EL}$ is equal to the solution to the estimating equations $\sum_{i=1}^{N}g(x_i,\theta) = 0$. When $r > q$, the empirical likelihood method allows us to deal with the combination of pieces of information about θ. However, computational issues may arise to obtain $\hat{\beta}_{EL}$.

Define correlation matrix $R_F(\alpha)$ as a toeplitz matrix with $(n-1)$-dimensional correlation parameter vector $\alpha = (\alpha_1, \ldots, \alpha_{n-1})^T$. That is, the jth off the diagonal element in $R_F(\alpha)$ is α_j. Therefore, $R_F(\alpha)$ is a more general structure, and the most commonly used correlation structures, independence, exchangeable, AR(1), and MA(1) are all embedded in $R_F(\alpha)$. We define the GEE model with $R_F(\alpha)$ as the "full model". Furthermore, unbiased estimating functions can be constructed for β and α,

$$
g(Y_i, X_i, \beta, \alpha, R_F) = \begin{pmatrix} D_i^T A_i^{-1/2} R_F^{-1}(\alpha) A_i^{-1/2} S_i \\ \sum_{k=1}^{n-1} e_{ij} e_{ik+1} - \alpha_1 \hat{\phi}(n-1-p/N) \\ \vdots \\ \sum_{k=1}^{1} e_{ij} e_{ik+n-1} - \alpha_{n-1} \hat{\phi}(1-p/N) \end{pmatrix},
$$

where $e_{ij} = (Y_{ij} - \mu_{ij})/\sqrt{v(\mu_{ij})}$. The empirical likelihood ratio based on $g(Y_i, X_i, \beta, \alpha, R_F)$ is

$$
\mathscr{R}^F(\beta, \alpha) = \sup \left\{ \prod_{i=1}^{N} N p_i; 0 \le p_i \le 1, \sum_{i=1}^{N} p_i = 1, \sum_{i=1}^{N} p_i g(Y_i, X_i, \beta, \alpha, R_F) = 0 \right\}. \quad (5.8)
$$

Note that this empirical likelihood is built at a very weak assumption, such as the stationarity assumption of the underlying correlation structure. $\mathscr{R}^F(\beta, \alpha)$ serves as a unified measure that can be applied to most of the competing GEE models (Chen and Lazar, 2012).

To avoid the computational issues associated with the empirical likelihood, Chen and Lazar (2012) proposed calculating $\mathscr{R}^F(\beta, \alpha)$ via giving β and α. If $n = 3$, the correlation matrix $R_F(\alpha)$ is parameterized by $\alpha^T = (\alpha_1, \alpha_2)$. Suppose that there are three correlation structure candidates: exchangeable (R_E), AR(1) (R_A), and toeplitz (R_F), and corresponding estimators for β and α are $(\hat{\beta}_E, \hat{\alpha}_E)$, $(\hat{\beta}_A, \hat{\alpha}_A)$, and $(\hat{\beta}_F, \hat{\alpha}_1^F, \hat{\alpha}_2^F)$, which are obtained with one of the three working correlation structures based on the same data. The corresponding empirical likelihood ratios are $\mathscr{R}^F(\hat{\beta}_E, \hat{\alpha}_E, \hat{\alpha}_E)$, $\mathscr{R}^F(\hat{\beta}_A, \hat{\alpha}_A, \hat{\alpha}_A)$, and $\mathscr{R}^F(\hat{\beta}_F, \hat{\alpha}_1^F, \hat{\alpha}_2^F)$. It is easy to see that $\mathscr{R}^F(\hat{\beta}_F, \hat{\alpha}_1^F, \hat{\alpha}_2^F)$ is equal to one, and the other models are smaller than one. Therefore, different competing structures embedded in the general correlation structure $R_F(\alpha)$ have different values, which can be used to compare the different competing structures and then select the best one.

Chen and Lazer (2012) modified the AIC and the BIC by substituting the empirical likelihood for the parametric likelihood and gave empirical likelihood versions of the AIC and the BIC:

$$
\text{EAIC} = -2 \log \mathscr{R}^F(\hat{\theta}) + 2 \dim(\theta),
$$

$$
\text{EBIC} = -2 \log \mathscr{R}^F(\hat{\theta}) + \log(m) \dim(\theta),
$$

where $\hat{\theta}_w = (\hat{\beta}_w^T, \hat{\alpha}_w^T)^T$. It is worth mentioning that, when the EAIC and the EBIC are calculated, $\hat{\beta}_w$ is the GEE estimator with the working correlation structure R_w, and $\hat{\alpha}$ is the method of moments estimator of α given $\hat{\beta}_w$ and R_w. Note that the EAIC and the EBIC contain $p + n - 1$ unknown parameters even for an exchangeable and AR(1) correlation matrix. Therefore, if sample size N is small and n is large, the performance of the EAIC and the EBIC is affected.

5.4   Examples

In this section, we illustrate the mentioned criteria for covariates selection and correlation structure selection by some real datasets. The R codes are also provided.

5.4.1   Examples for Variable Selection

In this subsection, we discuss the criteria for covariate selection under assumption that the variance function and correlation structure are given.

Example 1. Hormone data

The hormone data were introduced in **Chapter 1.1 Example 3**. The response variable is the log-transformed progesterone level. In addition to age and body mass index (BMI) covariates, we also consider the time effect. To account for highly non-linear, we utilize a polynomial to fit the time effect. We use the QIC, the EQIC, the GAIC, and the GBIC to select the degree of the polynomial. The results are presented in Table 5.1. As we can see that the QIC selects the quartic polynomial under two working correlation structures. However, the EQIC and the GAIC both choose the quintic polynomial. The GBIC chooses the quartic polynomial under the independence correlation structure but selects the quintic polynomial under the exchangeable correlation structure. According to Pan (2001), the QIC, the GAIC, and the GBIC with the independence correlation structure perform better than other cases. Furthermore, the values of the EQIC, the GAIC, and the GBIC for the quartic and quintic polynomials are close. In general, an underfitting model is more serious than an over-fitting model. Hence, we choose the quintic polynomial which is more conservative.

5.4.2   Examples for Correlation Structure Selection

In this subsection, we discuss the criteria for correlation structure selection under assumption that the variance function and covariates are given.

Example 2. Madras longitudinal schizophrenia study (binary data)

This study has been introduced in **Chapter 2**. To investigate the course of positive and negative psychiatric symptoms over the first year after initial hospitalization for schizophrenia, the response variable Y is binary, indicating the presence of thought disorder. The covariates are (a) Month: duration since hospitalization (in months), (b) Age: age of patient at the onset of symptom (1 represents Age <20; 0 otherwise), (c) Gender: gender of patient (1 =female; 0 =male), (d) Month×Age is the interaction term between variables Month and Age, (e) Month×Gender is the interaction term between variables Month and Gender, and (f) Age×Gender is the interaction term

Table 5.1 *The values of the criteria QIC, EQIC, GAIC, and GBIC under two different working correlation structures (CS) for Example 1.*

CS	Covariates	QIC	EQIC	GAIC	GBIC
	age,bmi, time	781.25	2258.19	2442.94	2447.52
	age, bmi,time, time^2	772.76	2233.12	2428.36	2434.47
Independence	age, bmi,time, time^2,time^3	670.72	1951.60	2250.88	2258.51
	age, bmi,time, time^2, time^3,time^4	**661.67**	1922.09	2232.42	**2241.57**
	age, bmi,time, time^2, time^3,time^4,time^5	663.27	**1921.63**	**2232.35**	2243.04
	covariates	QIC	EQIC	GAIC	GBIC
	age,bmi, time	784.12	2261.78	2170.26	2174.84
	age, bmi,time, time^2	775.67	2236.77	2147.46	2153.56
Exchangeable	age, bmi,time, time^2,time^3	673.45	1955.28	1874.48	1882.11
	age, bmi,time, time^2, time^3,time^4	**664.40**	1925.79	1843.43	1852.59
	age, bmi,time, time^2, time^3,time^4,time^5	666.32	**1925.65**	**1841.44**	**1852.12**

between Age and Gender. We consider the logistic regression.

$$\text{logit}(\mu_{it}) = \beta_0 + \beta_1 \text{Month}_{ik} + \beta_2 \text{Age}_i + \beta_3 \text{Gender}_i + \beta_4 \text{Month}_{ik} * \text{Age}_i$$
$$+ \beta_5 \text{Month}_{ik} * \text{Gender}_i + \beta_6 \text{Age}_i * \text{Gender}_i, \quad k = 1, \ldots, n_i; \quad i = 1, \ldots, 86.$$

We utilize the function *geeglm* in the *geepack* package to estimate parameters in the model. The parameter estimates obtained from the GEE with the independence, exchangeable, and AR(1) working correlation matrix are listed in Table5.2. The standard errors (Std.err) of the parameter estimators and P-values are also given.

When the working correlation structure is independence, the intercept and Month have significant positive effects on the probability of the presence of thought disorder. However, when the working correlation structure is exchangeable, besides the intercept and Month, the interaction term between Age and Gender are also significant, which indicates that the age has significantly different impact for female and male. The correlation coefficient estimate is 0.258, which indicates the correlation among the observations is weak. When the correlation structure is AR(1), the important covariates is the same as those under the independence working correlation strucutre. As we can see that the parameter estimates based on three different correlation structures are close, but the standard errors of the parameter estimators under the AR(1) correlation structure are smallest.

We next use the criteria to select a correlation structure for the GEE. The values of the criteria are presented in Table 5.3. The results indicate that all the criteria select the AR(1) correlation structure except the QIC. The QIC$_H$ and the CIC$_H$ correspond to the QIC and the CIC in which $\hat{\Omega}_I$ is evaluated via the GEE estimates based on the independence correlation structure.

Table 5.2 *The results obtained by the GEE with independence (IN), exchangeable (EX), and AR(1) correlation structures in Example 2.*

	Independence correlation matrix			
	Estimate	Std.err	Wald	Pr(>|W|)
Intercept	0.7969	0.3180	6.2812	0.0122
MONTH	-0.2627	0.0578	20.6750	0.0000
AGE	0.1967	0.6383	0.0950	0.7580
GENDER	-0.7200	0.4643	2.4052	0.1209
MONTH.AGE	-0.1006	0.0991	1.0311	0.3099
MONTH.GENDER	-0.1274	0.1076	1.4029	0.2362
AGE.GENDER	1.0495	0.7170	2.1425	0.1433
	Exchangeable correlation matrix			
	Estimate	Std.err	Wald	Pr(>|W|)
(Intercept)	0.8518	0.3232	6.9451	0.0084
MONTH	-0.2779	0.0613	20.5433	0.0000
AGE	0.1369	0.6684	0.0419	0.8378
GENDER	-1.2709	0.7256	3.0679	0.0799
MONTH.AGE	-0.0692	0.1007	0.4722	0.4920
MONTH.GENDER	-0.1643	0.1071	2.3555	0.1248
AGE.GENDER	1.7727	0.8968	3.9075	0.0481
	AR(1) correlation matrix			
	Estimate	Std.err	Wald	Pr(>|W|)
(Intercept)	0.6988	0.3076	5.1593	0.0231
MONTH	-0.2409	0.0538	20.0577	0.0000
AGE	0.0083	0.5962	0.0002	0.9889
GENDER	-0.4578	0.4481	1.0436	0.3070
MONTH.AGE	-0.0643	0.0894	0.5177	0.4718
MONTH.GENDER	-0.1771	0.0979	3.2735	0.0704
AGE.GENDER	1.0809	0.7136	2.2943	0.1298

In the statistical software R, we can first use the *geeglm* function in the *geepack* to obtain the parameter estimates, and then use the function *QIC* in the packages

Table 5.3 *The values of the criteria under independence (IN), exchangeable (EX), and AR(1) correlation structures in Example 2.*

	IN	EX	AR(1)
QIC	**954.12**	971.91	954.85
QIC_H	949.46	972.28	**948.88**
CIC	23.09	25.79	**22.69**
CIC_H	20.75	25.97	**19.71**
GAIC	882.53	1069.70	**477.69**
GBIC	899.71	1089.34	**497.32**
EAIC	2932.02	2782.40	**243.61**
EBIC	2949.20	2802.03	**263.24**

Table 5.4 *The results are obtained via the QIC function in MESS package.*

	QIC	QICu	Quasi Lik	CIC	params	QICC
IN	949.46	921.95	-453.98	20.75	7.00	949.58
EX	940.81	934.34	-460.17	10.24	7.00	940.97
AR	923.25	923.46	-454.73	6.90	7.00	923.41

geepack and *MESS* to calculate the QIC and the CIC. The specific results are as follows (Table 5.4).

In the output results, QIC and CIC correspond to the values of the QIC_H and the CIC_H, respectively, and QICu corresponds to the value of the QICu.

It is worth noting that the values of QIC and CIC obtained via *QIC* are different with those given in Table 5.3 except those under the independence working correlation matrix, because $\Omega_R = \sum_{i=1}^{N} D_i^T V_i^{-1} D_i$ is used in the *QIC* function instead of Ω_I.

Example 3. Progabide study

The Progbide study has been introduced in **Chapter 2**. The response variable Y is counted response, indicating the number of epileptic seizures of each patient. A log-linear regression model is considered:

$$\log(\mu_{ij}) = \log(t_{ij}) + \beta_0 + \beta_1 base_{ij} + \beta_2 trt_i$$
$$+ \beta_3 \log(age_{ij}) + \beta_4 trt_i \times base_{ij},$$

where $j = 0, \ldots, 4; i = 1, \ldots, 59$, and $\log(t_{ij})$ is an offset to take account of different interval lengths with $t_{ij} = 8$ if $j = 0$ and $t_{ij} = 2$ if $j = 1, 2, 3, 4$. The covariates $base_{ij} = 0$ if baseline, otherwise $base_{ij} = 1$; $trt_i = 1$ if the ith patient is in the progabide group, $trt = 0$ otherwise; age indicates the patients' age; the last term is the interaction effect between treatment and the baseline indicator. Therefore, the parameter β_1 is the logarithm of the ratio of the average seizure rate after treatment to before treatment for the placebo group. The parameter β_2 represents the difference in the logarithm of the average seizure rate at the baseline between the progabide and the placebo groups. The parameter β_4 is of interest and indicates the difference in the logarithm of the post to pre-treatment ration between the progabide and the placebo groups.

To account for the over-dispersion which is not described by the Poisson distribution, we assume that $var(y_{ij}) = \phi E(y_{ij})$ with $\phi > 1$. The function *summary* in R is used to produce summaries of the results of various model fitting functions. The function *update* is used to update and refit a model. The results for the parameter estimates obtained from the GEE with the independence, exchangeable, and AR(1) working correlation structures are given in Table 5.5.

The parameter estimates obtained from the GEE with the three correlation structures are similar. The over-dispersion parameter ϕ is significantly larger than 1. Parameter estimate of β_4 is negative, which indicates that the progabide can reduce the number of epileptic seizure counts. The estimate of the correlation coefficient is larger than 0.7, which indicates there exists strong correlations. The results obtained

Table 5.5 *The results obtained by the GEE with independence (IN), exchangeable (EX), and AR(1) correlation structures in Example 3.*

	Independence correlation matrix			
	Estimate	Std.err	Wald	Pr(>|W|)
Intercept	2.8887	1.6027	3.2487	0.0715
base	0.1118	0.1159	0.9306	0.3347
trt	0.0090	0.2089	0.0018	0.9658
log(age)	-0.4618	0.4852	0.9059	0.3412
base:trt	-0.1047	0.2134	0.2407	0.6237
	Exchangeable correlation matrix			
	Estimate	Std.err	Wald	Pr(>|W|)
Intercept	3.9705	1.2129	10.7160	0.0011
base	0.1118	0.1159	0.9306	0.3347
trt	-0.0086	0.2147	0.0016	0.9681
log(age)	-0.7876	0.3649	4.6583	0.0309
base:trt	-0.1047	0.2134	0.2407	0.6237
	AR(1) correlation matrix			
	Estimate	Std.err	Wald	Pr(>|W|)
Intercept	4.5580	1.3991	10.6127	0.0011
base	0.1561	0.1137	1.8843	0.1699
trt	-0.0312	0.2093	0.0223	0.8814
log(age)	-0.9765	0.4186	5.4409	0.0197
base:trt	-0.1315	0.2667	0.2431	0.6220

under the exchangeable and AR(1) correlation matrices indicate that intercept and log(age) are significant.

Patient number 207 appears to be very unusual (*id* = 49 in the dataset). He (or she) had an extremely high seizure count (151 seizures in eight weeks) at baseline and his count doubled after treatment (302 seizures in eight weeks). The GEE method is sensitive to outliers, and hence we rerun the results after removing the outliers and present the new results in Table 5.6.

The results are different with and without the outliers. After removing the outliers, the interactive term between base and treatment is significant, which indicates the difference in the logarithm of the post to pre-treatment ration between the progabide and the placebo groups are significant.

Next, we use the criteria to specify the correlation structure. We also calculate the criteria with and without the 49th patient. The results are presented in Table 5.7. When the data contain the outliers, all the criteria select the exchangeable correlation structure. However, when the data exclude the outliers, the QIC and QIC$_H$ choose the independence correlation structure, and the CIC chooses the AR(1) correlation structure, which indicate that the QIC and the CIC are sensitive to outliers. We believe that the exchangeable correlation structure is more reliable. We also calculate the sample correlation matrix (see Table 5.8), which or that is close to the exchangeable correlation structure.

Table 5.6 *The results obtained by the GEE with independence (IN), exchangeable (EX), and AR(1) correlation structures when removing the outliers in Example 3.*

	Independence correlation matrix			
	Estimate	Std.err	Wald	Pr(>|W|)
Intercept	1.6165	1.3153	1.5104	0.2191
base	0.1118	0.1159	0.9306	0.3347
trt	-0.1089	0.1909	0.3253	0.5684
log(age)	-0.0804	0.3966	0.0411	0.8394
base:trt	-0.3024	0.1711	3.1248	0.0771
	Exchangeable correlation matrix			
Intercept	3.2250	1.2461	6.6977	0.0097
base	0.1118	0.1159	0.9306	0.3347
trt	-0.1265	0.1884	0.4505	0.5021
log(age)	-0.5629	0.3753	2.2497	0.1336
base:trt	-0.3024	0.1711	3.1248	0.0771
	AR(1) correlation matrix			
Intercept	4.1477	1.3083	10.0503	0.0015
base	0.1521	0.1114	1.8654	0.1720
trt	-0.1143	0.1943	0.3457	0.5566
log(age)	-0.8515	0.3918	4.7236	0.0298
base:trt	-0.4018	0.1757	5.2314	0.0222

Table 5.7 *The results of the criteria for the data with and without the outliers in Example 3.*

	Complete data			Outlier deleted		
	IN	EX	AR	IN	EX	AR
QIC	-695.76	**-701.26**	-678.12	**-1062.05**	-1014.30	-938.22
QIC$_H$	-695.76	**-701.42**	-676.64	**-1062.05**	-1013.18	-934.95
CIC	11.79	**10.62**	11.00	11.03	10.32	**9.47**
CIC$_H$	11.79	**10.54**	11.73	11.03	**10.88**	11.11
C(R)	13.57	**2.18**	12.38	9.61	**2.74**	10.66
GAIC	2409.43	**2142.65**	2256.92	2158.03	**2017.64**	2098.56
GBIC	2419.81	**2155.12**	2269.39	2168.33	**2030.00**	2110.92
EAIC	142.68	**13.33**	60.13	1894.41	**17.74**	26.56
EBIC	153.07	**25.79**	72.59	1904.71	**30.10**	38.92

Table 5.8 *The correlation matrix of the data with and without the outliers in Example 3.*

	Complete data					Outlier deleted			
0	1.00					1.00			
1	0.79	1.00				0.68	1.00		
2	0.83	0.87	1.00			0.73	0.69	1.00	
3	0.67	0.74	0.80	1.00		0.49	0.54	0.67	1.00
4	0.84	0.89	0.89	0.82		0.75	0.72	0.76	0.71

Chapter 6

Robust Approaches

6.1 Introduction

The GEE method is robust against the misspecification of correlation structures, but is sensitive to outliers because it is essentially a generalized weighted least squares approach (Jung and Ying, 2003; Wang and Zhao, 2008). In this chapter, we will introduce several robust methods for parameter estimation in analysis of longitudinal data. In §6.2, we introduce the rank-based method, which is distribution-free, robust, and highly efficient (Hettmansperger, 1984). In §6.3, we will introduce the quantile regression, which gives a global assessment about covariate effects on the distribution of the response variable, provides complete description of the distribution, and is robust against outliers (Koenker and Bassett, 1978). In §6.4, we will introduce other methods based on the Huber's function and exponential square loss function, which is not only robust against outliers in response variable, but also robust against outliers in covariates.

6.2 Rank-based Method

Rank-based methods are robust and highly efficient (Hettmansperger, 1984). Jung and Ying (2003) extended the Wilcoxon-Man-Whitney rank statistic to the longitudinal data under the independence assumption. Their method is simple and also has desirable properties. In this section, we will introduce several rank-based methods for parameter estimation in linear regression models.

Let y_{ik} be the $n_i \times 1$ vector of responses for $i = 1, \ldots, N$. Then the model for y_{ik} is

$$y_{ik} = X_{ik}^{\mathrm{T}} \beta + \epsilon_{ik}, \quad k = 1, \ldots, n_i, \quad i = 1, \ldots, N, \tag{6.1}$$

where β is the parameter vector corresponding to the covariate vector X_{ik} of dimension p. Assume that the median of $\epsilon_{ik} - \epsilon_{jl}$ is zero. To avoid complication caused by ties, we assume that error terms ϵ_{ik} are continuous variables. Define residuals $e_{ik}(\hat{\beta}) = y_{ik} - X_{ik}^{\mathrm{T}} \hat{\beta}$, where $\hat{\beta}$ is a given consistent estimator of β.

6.2.1 An Independence Working Model

Let M be the total number of observations and $\bar{X} = M^{-1} \sum_{i=1}^{N} \sum_{k=1}^{n_i} x_{ik}$. The corresponding ranks of $e_{ik}(\beta)$ are $r_{ik}(\beta) = \sum_{j=1}^{N} \sum_{l=1}^{n_j} I(e_{jl} \leq e_{ik})$. Jung and Ying (2003)

DOI: 10.1201/9781315153636-6

proposed estimating β using the following estimating functions:

$$U_I(\beta) = M^{-1} \sum_{i=1}^{N} \sum_{k=1}^{n_i} (X_{ik} - \bar{X}) r_{ik}(\beta). \tag{6.2}$$

Let $\hat{\beta}_I$ be the resultant estimator from (6.2), which can also be obtained by minimizing the following loss function

$$L(\beta) = M^{-2} \sum_{i=1}^{N} \sum_{k=1}^{n_i} \sum_{j=1}^{N} \sum_{l=1}^{n_j} |e_{ik}(\beta) - e_{jl}(\beta)|. \tag{6.3}$$

To make inferences about the regression coefficients of model (6.1), such as testing the null hypothesis $H_0 : \beta = \beta_0$, Jung and Ying (2003) proposed a test statistic $N U_I^{T}(\beta) V^{-1} U_I(\beta)$, which can bypass density estimation and bandwidth selection, and V is a covariance matrix of $N^{-1/2} U_I(\beta)$. A consistent estimator of V is $\hat{V} = N^{-1} \sum_{i=1}^{N} \hat{\xi}_i \hat{\xi}_i^{T}$, where

$$\hat{\xi}_i = \sum_{k=1}^{n_i} \left[(X_{ik} - \bar{X})\{N^{-1} r_{ik}(\hat{\beta}_I) - 1/2\} - N^{-1} \sum_{i'=1}^{N} \sum_{k'=1}^{n_{i'}} (X_{i'k'} - \bar{X}) I(e_{i'k'}(\hat{\beta}_I) \le e_{ik}(\hat{\beta}_I)) \right],$$

Under the null hypothesis, the test statistic approximately has a χ^2 distribution with p degrees of freedom. The null hypothesis will be rejected for a large observed value of the test statistic.

The function (6.3) is based on the independence working model assumption, thus the efficiency of $\hat{\beta}_I$ may be improved by taking account of the correlations and the impacts of varying cluster sizes. To this end, Wang and Zhao (2008) introduced a weighted estimating functions.

6.2.2 A Weighted Method

To avoid modeling the correlations, we can resample one observation from each subject, and then apply the rank method to the N independent observations, $(y_i), 1 \le i \le N$. If the corresponding residuals for the ith subject are $(e_{(i)})$, the dispersion function is

$$\sum_{i=1}^{N} \sum_{j \ne i}^{N} |e_{(i)} - e_{(j)}|.$$

To make use of all the observations, we would repeat this resampling process many times. Conditional on sampling one observation from each cluster, the probability of y_{ik} being sampled is $1/n_i$. Therefore, the limiting dispersion function is

$$L_w(\beta) = \frac{1}{M^2} \sum_{i=1}^{N} \sum_{j \ne i}^{N} \frac{1}{n_i} \frac{1}{n_j} \sum_{k=1}^{n_i} \sum_{l=1}^{n_j} |e_{(i)} - e_{(j)}|.$$

This motivates Wang and Zhao (2008) to consider the following weighted loss function for estimation of β,

$$L_{wz}(\beta) = \frac{1}{M^2} \sum_{i=1}^{N} \sum_{j \neq i}^{N} \sum_{k=1}^{n_i} \sum_{l=1}^{n_j} \omega_i \omega_j |e_{ik} - e_{jl}|,$$

where ω_i and ω_j are weighted functions. The corresponding quasi-score functions for β are

$$U_{wz}(\beta) = \frac{1}{M^2} \sum_{i=1}^{N} \sum_{j \neq i}^{N} \sum_{k=1}^{n_i} \sum_{l=1}^{n_j} \omega_i \omega_j (x_{ik} - x_{jl}) I(e_{jl} \leq e_{ik}), \tag{6.4}$$

For the weighted function ω_i, it seems that sensible choices include (i) $\omega_i = 1$, (ii) $\omega_i = 1/n_i$, and (iii) $\omega_i = 1/(1 + (n_i - 1)\bar{\rho})$, where $\bar{\rho}$ is the "average" within correlation and can be estimated by

$$\hat{\rho}_0 = \frac{\sum_{i=1}^{N} \sum_{k \neq l}^{n_i} (r_{ik} - \bar{r})(r_{il} - \bar{r})}{\sum_{i=1}^{N} \sum_{k=1}^{n_i} (n_i - 1)(r_{ik} - \bar{r})^2},$$

where $\bar{r}$ is the mean of the ranks r_{ik} for $k = 1 \ldots, n_i$ and $i = 1, \ldots, N$. When the within correlation is weak, it can be ignored in the analysis, and $\omega_i = 1$ is corresponding to the classical rank regression.

Let $f_{ik}(\cdot)$ and $F_{ik}(\cdot)$ be the probability density function and the cumulative distribution function of ϵ_{ik}, for $k = 1, \ldots, n_i$ and $i = 1, \ldots, N$. Let $V_w = \lim_{M \to \infty} \sum_{i=1}^{N} E(\eta_i \eta_i^T)$, where

$$\eta_i = \sum_{j \neq i}^{N} \omega_i \omega_j \sum_{k=1}^{n_i} \sum_{l=1}^{n_l} (x_{ik} - x_{jl}) \left\{ F_{jl}(\epsilon_{ik}) - \frac{1}{2} \right\}.$$

Furthermore, let

$$D_N = \frac{2}{M^2} \sum_{i=1}^{N} \sum_{j \neq i}^{N} \omega_i \omega_j \sum_{k=1}^{n_i} \sum_{l=1}^{n_l} (x_{ik} - x_{jl})(x_{ik} - x_{jl})^T \int f_{ik} dF_{ik}.$$

Assume that the limiting matrix of D_N as $N \to \infty$ is D_w. If n_i is bounded and both V_w and D_w are positive definite, then the resultant estimator $\hat{\beta}_{wz}$ from (6.4) has the following asymptotic result: $\sqrt{M}(\hat{\beta}_{wz} - \beta)$ converges in distribution to $N(0, \Lambda_w)$, where $\Lambda_w = D_w^{-1} V_w D_w^{-1}$. The covariance matrix V_w can be replaced with a consistent estimator $\hat{V}_w = 16/M^3 \sum_{i=1}^{N} \hat{\eta}_i \hat{\eta}_i^T$, in which

$$\hat{\eta}_i = \sum_{j \neq i}^{N} \omega_i \omega_j \sum_{k=1}^{n_i} \sum_{l=1}^{n_l} (x_{ik} - x_{jl}) \left\{ I(e_{ik} \geq e_{jl}) - \frac{1}{2} \right\}.$$

The proposed weighted rank method is simple and effective because it utilizes only the weight function to incorporate correlations and cluster sizes for efficiency gain. However, when the design is balanced, this method is equivalent to the Jung and Ying's method. Furthermore, this method reduces the efficiency of parameter estimators because of only weighting at cluster levels, in which the covariate varies within each subject (Neuhaus and Kalbfleish, 1998).

6.2.3 Combined Method

Decompose $L(\beta)$ given in (6.3) into two parts, as

$$L(\beta) = M^{-2} \sum_{i=1}^{N} \sum_{j \neq i}^{N} \sum_{k=1}^{n_i} \sum_{l=1}^{n_j} |e_{ik}(\beta) - e_{jl}(\beta)| + M^{-2} \sum_{i=1}^{N} \sum_{k=1}^{n_i} \sum_{l=1}^{n_i} |e_{ik}(\beta) - e_{il}(\beta)|$$

$$= L_B(\beta) + M^{-1} L_W(\beta),$$

where $L_B(\beta)$ and $L_W(\beta)$ stand for between- and within-subject loss functions. As M tends to infinity, $L(\beta) \simeq L_B(\beta) + O_p(1)$, and thus the information contained in the within-subject loss function $L_W(\beta)$ is ignored. Therefore, Wang and Zhu (2006) proposed minimizing $L_B(\beta)$ and $L_W(\beta)$ separately and then combining the two aforementioned estimators in some optimal sense.

The estimating functions based on the between- and within-subject ranks $L_B(\beta)$ and $L_W(\beta)$ can be obtained:

$$S_B(\beta) = \frac{1}{M^2} \sum_{i=1}^{N} \sum_{j \neq i}^{N} \sum_{k=1}^{n_i} \sum_{l=1}^{n_j} (X_{ik} - X_{jl}) I(e_{jl} \leq e_{ik}), \tag{6.5}$$

$$S_W(\beta) = \frac{1}{M} \sum_{i=1}^{N} \sum_{k \neq l}^{n_i} (X_{ik} - X_{il}) I(e_{il} \leq e_{ik}). \tag{6.6}$$

Because the median of $e_{ik} - e_{jl}$ is zero, (6.5) and (6.6) are unbiased and can be used to estimate β. Suppose that $\hat{\beta}_B$ and $\hat{\beta}_W$ are estimators derived from (6.5) and (6.6), respectively. It seems natural to combine these two estimators to obtain a more efficient estimator (Heyde, 1987). Let $X_B = M^{-2} \sum_{i=1}^{N} \sum_{j \neq i}^{N} \sum_{k=1}^{n_i} \sum_{l=1}^{n_j} (X_{ik} - X_{jl})(X_{ik} - X_{jl})^{\mathrm{T}}$ and $X_W = M^{-1} \sum_{i=1}^{N} \sum_{k=1}^{n_i} \sum_{l=1}^{n_i} (X_{ik} - X_{il})(X_{ik} - X_{il})^{\mathrm{T}}$. Wang and Zhu (2006) proposed finding an optimal estimator of β from an optimal linear combination of the estimating functions for β:

$$S_C(\beta) = (X_B, X_W) \Sigma^{-1} \begin{pmatrix} S_B(\beta) \\ S_W(\beta) \end{pmatrix} \tag{6.7}$$

where Σ is the covariance matrix of $S_B(\beta)$ and $S_W(\beta)$. Because Σ is unknown, Wang and Zhu (2006) proposed using resampling method to bypass density estimation. Suppose that $\{z_i\}_{i=1}^{N}$ are independently sampled from the binomial distribution $B(N, 1/N)$. The bootstrap method of resampling the subjects with replacement leads to the following perturbed estimating functions:

$$\tilde{S}_B(\beta) = \frac{1}{M^2} \sum_{i=1}^{N} \sum_{j \neq i}^{N} \sum_{k=1}^{n_i} \sum_{l=1}^{n_j} z_i z_j (X_{ik} - X_{jl}) I(e_{jl} \leq e_{ik}),$$

$$\tilde{S}_W(\beta) = \frac{1}{M} \sum_{i=1}^{N} \sum_{k \neq l}^{n_i} z_i (X_{ik} - X_{il}) I(e_{il} \leq e_{ik}).$$

The covariance matrix of $\tilde{S}_B$ and $\tilde{S}_W$ can be used as an estimate of Σ.

It is worth noting that in the case of cluster-level covariate designs, in which the covariate has the same value for all the units in the same subject/cluster, the within-subject estimating functions are identically equal to zero, and hence the combined method is equivalent to the independence working model.

6.2.4 A Method Based on GEE

To improve the efficiency of estimators by incorporating the within correlations, Fu and Wang (2012) derived an optimal rank-based estimating function in terms of asymptotic variance of regression parameter estimators (Heyde, 1987) under the GEE framework.

Let $S_{ik} = M^{-1} \sum_{j=1}^{N} \sum_{l=1}^{n_j} \{I(e_{jl}(\beta) \leq e_{ik}(\beta)) - 1/2\}$ be the average rank of $e_{ik}(\beta)$. The indicator function $I(e_{jl}(\beta) \leq e_{ik}(\beta))$ minus $1/2$ is to make sure S_{ik} is unbiased. Let $S_i(\beta) = (S_{i1}, \ldots, S_{in_i})^T$ and $S(\beta) = (S_1(\beta)^T, \ldots, S_N(\beta)^T)^T$. Because the summation of $S(\beta)$ is a constant $M(M+1)/2$, the covariance matrix V_G of $S(\beta)$ is singular. Therefore, Fu and Wang (2012) proposed randomly selecting $M-1$ elements of $S(\beta)$ to estimate β. We will still use $S(\beta)$ and V_G to indicate the $M-1$ average ranks, and their covariance matrix, respectively.

According to Heyde (1987) and Durairajan (1992), Fu and Wang (2012) proposed an optimal combination of all the elements of $S(\beta)$ which takes the form of

$$S_O(\beta) = \bar{D}^T V_G^{-1} S(\beta), \tag{6.8}$$

where $\bar{D} = \{\bar{D}_{ik}, k = 1, \ldots, n_i; i = 1, \ldots, N\}$ are derivatives of expected values of $S(\beta)$, and $\bar{D}_{ik}^T = M^{-1} \sum_{j=1}^{N} \sum_{l=1}^{n_j} (X_{jl} - x_{ik}) f_{ikjl}(0)$, where $f_{ikjl}(\cdot)$ is the density function of $\epsilon_{ik} - \epsilon_{jl}$. Note that $\bar{D}_{ik}$ involves the unknown density function $f_{ikjl}(\cdot)$. To bypass estimate $f_{ikjl}(0)$, Fu and Wang (2012) assumed that $f_{ikjl}(0)$ is a constant. Therefore, we can replace $\bar{D}_{ik}^T$ by $D_{ik}^T = \bar{x} - x_{ik}$ in the function (6.8).

Due to a variety of correlation types induced by combinations of the between- and within-subject ranks, it is even difficult to specify the correlation structure of $S(\beta)$. Under an exchangeable correlation structure of the within-subject residuals assumption, Fu and Wang (2012) derived V_G and used this matrix as a working matrix in (6.8). Let $\sigma_i^2 = \text{Var}(S_{ik})$, $\sigma_{ii} = \text{Cov}(S_{ik}, S_{il})$, $\sigma_{ij} = \text{Cov}(S_{ik}, S_{jl})$, $r_{1i} = \sum_{j \neq i}^{N} n_j$, $r_{2i} = \sum_{j \neq i}^{N} n_j^2$, and $r_{3ij} = n_i + n_j - 1$. Fu and Wang (2012) obtained

$$\sigma_i^2 = \{r_{1i} + (r_{2i} - r_{1i})\rho_3 + (r_{1i}^2 - r_{2i})\rho_4\}\tau_2^2 + (n_i - 1)\tau_1\{2r_{1i}\rho_2\tau_2 + [1 + (n_i - 2)\rho_1]\tau_1\}/M^2,$$

$$\sigma_{ii} = \{[r_{1i}\rho_3 + (r_{2i} - r_{1i})\rho_5 + (r_{1i}^2 - r_{2i})\rho_6]\tau_2^2 - 2r_{1i}\rho_2\tau_1\tau_2 - [(n_i - 2)\rho_1 + 1]\tau_1^2\}/M^2,$$

$$\sigma_{ij} = \{[(r_{2i} - n_j^2)(r_{1i} - n_j)r_{3ij}]\rho_6 - 1 - (r_{1i} - n_j)\rho_4 - (r_{3ij} - 1)\rho_3 - (n_i n_j - r_{3ij})\rho_5]\tau_2^2/M^2,$$

where $\tau_1^2 = \text{var}\{I(e_{ik} \leq e_{il})\}$, $\tau_2^2 = \text{var}\{I(e_{ik} \leq e_{jl})\}$, and $\rho_1, \ldots, \rho_6$ are six types of correlation coefficients among $\{S_{ik}(\beta), k = 1, \ldots, n_i; i = 1, \ldots, N\}$ and given as follows

$$\rho_1 = corr\{I(e_{ik} \leq e_{il}), I(e_{ik} \leq e_{il'})\}, \quad \rho_2 = corr\{I(e_{ik} \leq e_{il}), I(e_{ik} \leq e_{rs})\},$$

$$\rho_3 = corr\{I(e_{ik} \leq e_{jl}), I(e_{ik} \leq e_{jl'})\}, \quad \rho_4 = corr\{I(e_{ik} \leq e_{jl}), I(e_{ik} \leq e_{rs})\},$$

$$\rho_5 = corr\{I(e_{ik} \leq e_{jl}), I(e_{ik'} \leq e_{jl'})\}, \quad \rho_6 = corr\{I(e_{ik} \leq e_{jl}), I(e_{ik'} \leq e_{rs})\}.$$

These six types correlation coefficients illustrate the correlations between the relative ordering of two different within correlated pairwise residuals. For example, ρ_1 indicates the correlation between the relative ordering of two pairwise residuals from the same subject, and ρ_6 indicates the correlation between the relative ordering of two different residuals from the same subject and another two distinct residuals from different subjects.

Note that if there are no within correlations, $\rho_5 = \rho_6 = 0$. If there exists correlations, and the errors have an exchangeable structure, then the pairwise residuals in ρ_1 and ρ_4 are identically distributed, hence $\rho_1 = \rho_4 = 1/3$. Furthermore, it can be seen that $\sigma_i^2 = O(1)$, $\sigma_{ii}^2 = O(1)$, and $\sigma_{ij} = O(M^{-1})$ for $i \neq j$, which implies that correlations between rank residuals from different subjects can be ignored as M tends to infinity. Utilizing the idea of the GEE, Fu and Wang (2012) proposed using a block diagonal matrix $\mathrm{diag}(V_1, \cdots, V_N)$ as a working covariance matrix for V, and obtained an estimate of β using estimating equations

$$S_G(\beta) = \sum_{i=1}^{N} D_i^{\mathrm{T}} V_i^{-1} S_i(\beta) = 0, \tag{6.9}$$

where $V_i^{-1} = (\sigma_i^2 - \sigma_{ii})^{-1} \{ I_{n_i} - \sigma_{ii} [\sigma_i^2 + (n_i - 1)\sigma_{ii}]^{-1} J_{n_i \times n_i} \}$ is the inverse matrix of $\mathrm{Cov}(S_i(\beta))$, in which I is an identity matrix, and J is a matrix with all elements being one. The equation (6.9) is the generalized estimating equations of $S_i(\beta)$. Let $\hat{\beta}_G$ be the resulting estimator from (6.9), then it can be shown that $\sqrt{N}(\hat{\beta}_G - \beta)$ is asymptotically normal with mean zero and covariance matrix

$$\Sigma_G = \lim_{N \to \infty} N \left(\sum_{i=1}^{N} D_i^{\mathrm{T}} V_i^{-1} \bar{D}_i \right)^{-1} \left(\sum_{i=1}^{N} D_i^{\mathrm{T}} V_i^{-1} \mathrm{Cov}\{S_i(\beta_0)\} V_i^{-1} D_i \right) \left(\sum_{i=1}^{N} \bar{D}_i^{\mathrm{T}} V_i^{-1} D_i \right)^{-1}.$$

To avoid estimating the joint density of the error terms, we use the resampling method to estimate Σ_G. Let $\{z_i\}_{i=1}^{N}$ be sampled from a distribution with unit mean and unit variance, and define the perturbed estimating equation

$$\tilde{S}_G(\beta) = \sum_{i=1}^{N} z_i D_i^{\mathrm{T}} V_i^{-1} S_i(\beta) = 0. \tag{6.10}$$

We can derive an estimate of β by solving the equation (6.10) for each sequence $\{z_i\}_{i=1}^{N}$. Therefore, an independent sequence of $\{z_i\}_{i=1}^{N}$ can result in many estimates of β. The covariance of the bootstrap estimates can be an estimate of Σ_G.

6.2.5 Pediatric Pain Tolerance Study

In this subsection, we will use a real data set to illustrate the rank-based methods. The data are from a pediatric pain tolerance study (Weiss et al., 1999) which contained a total of 64 children aged $8-10$ years old. The children attended four trials of keeping their arms in very cold water for as long as possible. The response variable is the log$_2$ of time in seconds that the children can tolerate keeping their arms immersed in cold water. Two trials were made during a first visit, and two more trials were made during a second visit two weeks later. According to whether focusing on the experiment or not, the children were classified as attenders or distracters based on their coping responses during the first two trials. Three baseline trials were given and then children were randomly assigned to one of three counseling interventions before the last trial, where advice was given to attend the experiment or to distract the experiment, or no intervention. This study aims at comparing the two baselines and examining whether treatment effects differ for attenders and distracters. There are missing data from six children because of arms in casts and similar reasons. We assume that the data are missing completely at random.

In Figure 6.1, it can be seen that the distract treatment appears to improve the performance of distracters. The attend treatment appears to decrease the pain tolerance time of the distracters. In Figure 6.2, girls have relatively longer pain tolerance time than boys over the whole trial period, and there are some outliers as pointed out by Weiss et al. (1998). Figure 6.3 indicates there exists strong correlations among the observations. Therefore, robust and efficient rank methods are desirable to analyze this dataset.

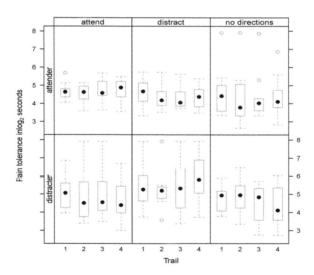

Figure 6.1 *The Boxplots of the time in* log$_2$ *seconds of pain tolerance in four trials in the pediatric pain study. Two row panels represent attender and distracter baseline groups, and three column panels represent three treatments.*

Let y_{ik} be the log$_2$ of pain tolerance time of the kth trial for the ith subject, and B be the baseline indicator taking 1 for attenders and 0 for distracters. The attend treatment is denoted by A, and the distract treatment is denoted by D, and no advice treatment is denoted by F. Let $A = 1$ for the attend treatment and $A = 0$ otherwise,

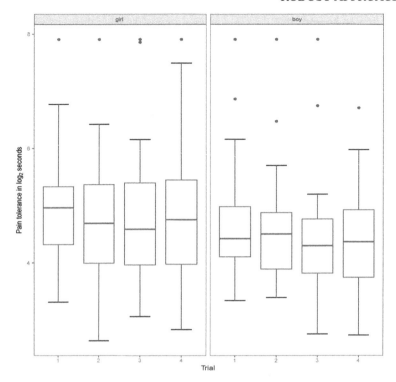

Figure 6.2 *The boxplots of the time in* $\log_2$ *seconds of pain tolerance are for girls and boys in four trails in the pediatric pain study.*

$D = 1$ for the distract treatment and $D = 0$ otherwise, and $F = 1$ for no intervention treatment and $F = 0$ otherwise. G is a gender indicator taking 1 for girl and 0 for boy. We assume that the data are missing completely at random and further consider the following model

$$y_{ik} = \beta_0 + \beta_1 B_i + (\beta_2 A_{ik} + \beta_3 D_{ik} + \beta_4 F_{ik}) * B_i$$
$$+ (\beta_5 A_{ik} + \beta_6 D_{ik} + \beta_7 F_{ik}) * (1 - B_i) + \beta_8 G_i + \epsilon_{ik},$$

where β_0 is the mean of the distracter group (baseline), β_1 indicates the difference between the attender and distracter groups. Parameters $(\beta_2, \beta_3, \beta_4)$ correspond the three treatment effects for the attender group, and $(\beta_5, \beta_6, \beta_7)$ correspond the three treatments for the distracter group. The standard errors for estimates obtained by the JY's method and Wang and Fu's method are based on 3000 resampling estimates from an exponential distribution with unit variance.

Parameter estimates and their standard errors obtained from different methods are given in Table 6.1. The results obtained via the GEE method depend on the selection of the working correlation matrix. Comparing to the GEE method, all the rank-based methods indicate that the distract treatment given to distracters will help distracters

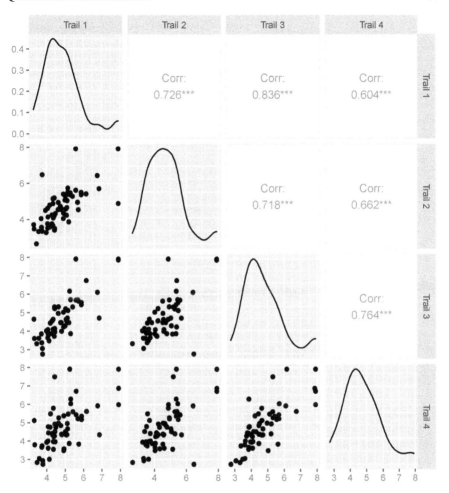

Figure 6.3 *Scatterplot of the time in* $\log_2$ *seconds of pain tolerance for four trails in the pediatric pain study. The diagonal plots show the densities of responses in the four Trails. The plots were produced by ggpairs in GGally.*

increase their pain tolerance (the estimate for β_6 is significant), which may have implications for medical treatments with painful procedures. In addition, except the GEE with an exchangeable working matrix, all the other methods indicate that the girls have much stronger pain tolerance than boys as Figure 6.2 indicates.

6.3 Quantile Regression

Various methods have been developed to evaluate covariate effects on the mean of a response variable in longitudinal data analysis (Liang and Zeger, 1986; Qu et al., 2000). To give a global assessment about covariate effects on the distribution

Table 6.1 *Parameter estimates and their standard errors (SE) for the pediatric pain tolerance study. JY: the method of Jung and Ying (2003); CM: the method of Wang and Zhu (2006); WZ: the method by Wang and Zhao (2008); FW: the method by Fu and Wang (2012); $GEE_{AR(1)}$: the GEE method with an independence working matrix; GEE_{EX}: the GEE method with an exchangeable working matrix; and $GEE_{AR(1)}$: the GEE method with an AR(1) working matrix; GEE_{UN}: the GEE method with an unstructure working matrix. * indicates that p-value is less than 0.05.*

	JY	CM	WZ	FW	GEE_{IN}	GEE_{EX}	$GEE_{AR(1)}$	GEE_{UN}
β_1	-0.57*	-0.47*	-0.57*	-0.59*	-0.60*	-0.59*	-0.71	-0.65*
(SE)	(0.18)	(0.24)	(0.21)	(0.18)	(0.24)	(0.25)	(0.24)	(0.24)
β_2	0.31	0.24	0.34	0.27	0.11	0.07	0.11	0.21
(SE)	(0.22)	(0.22)	(0.23)	(0.22)	(0.18)	(0.13)	(0.14)	(0.23)
β_3	-0.002	-0.06	0.03	0.15	0.04	0.04	0.13	-0.07
(SE)	(0.23)	(0.28)	(0.25)	(0.26)	(0.25)	(0.29)	(0.29)	(0.24)
β_4	-0.25	-0.11	-0.25	-0.38	-0.11	-0.11	-0.01	-0.16
(SE)	(0.33)	(0.21)	(0.26)	(0.26)	(0.12)	(0.13)	(0.13)	(0.25)
β_5	-0.36	-0.28	-0.39	-0.43*	-0.35*	-0.27*	-0.10	-0.39
(SE)	(0.28)	(0.20)	(0.26)	(0.19)	(0.16)	(0.13)	(0.22)	(0.27)
β_6	0.95*	0.87*	0.80*	1.07*	0.57	0.55	0.66	0.87*
(SE)	(0.43)	(0.36)	(0.42)	(0.37)	(0.30)	(0.35)	(0.31)	(0.35)
β_7	-0.69*	-0.60*	-0.71*	-0.54*	-0.48*	-0.31*	-0.38	-0.81*
(SE)	(0.33)	(0.27)	(0.35)	(0.25)	(0.20)	(0.13)	(0.16)	(0.31)
β_8	0.46*	0.40*	0.47*	0.59*	0.44	0.50*	0.59	0.43
(SE)	(0.17)	(0.22)	(0.21)	(0.16)	(0.24)	(0.23)	(0.21)	(0.23)

of the response variable, we will introduce a quantile regression model proposed by Koenker and Bassett (1978), which can characterize a particular point of a distribution and hence provide more complete description of the distribution. Furthermore, quantile regression is robust against outliers and does not require specifying any error distribution.

Assume that the $100\tau^{th}$ percentile of y_{ik} is $X_{ik}^{T}\beta_{\tau}$, where β_{τ} is a $p \times 1$ unknown parameter vector. The conditional quantile functions of the response y_{ik} is:

$$Q_{\tau}(y_{ik}|X_{ik}) = X_{ik}^{T}\beta_{\tau},$$

let $\epsilon_{ik} = y_{ik} - X_{ik}^{T}\beta_{\tau}$ be a continuous error term satisfying $p(\epsilon_{ik} \le 0) = \tau$ and with an unspecified density function $f_{ik}(\cdot)$. The median regression is obtained by taking $\tau = 0.5$.

6.3.1   An Independence Working Model

Let $s_{ik} = \tau - I(y_{ik} - X_{ik}^{T}\beta_{\tau} \le 0)$, where $I(\cdot)$ is an indicator function. Because $\epsilon_{i1}, \ldots, \epsilon_{in_i}$ are correlated, $s_{i1}, \ldots, s_{in_i}$ are also correlated for $i = 1, \ldots, N$. However, their correlation structure is more complex. Under the independence working model assumption, Chen et al. (2004) proposed using the following estimating functions to make inferences about β_τ:

$$U_\tau(\beta) = \sum_{i=1}^{N} \sum_{k=1}^{n_i} X_{ik} s_{ik}. \tag{6.11}$$

The resulting estimates $\hat{\beta}_{\tau I}$ from (6.11) can be also derived by minimizing the following loss function

$$L_\tau(\beta) = \sum_{i=1}^{N} \sum_{k=1}^{n_i} \rho_\tau(y_{ik} - X_{ik}^{\mathrm{T}}\beta), \tag{6.12}$$

where $\rho_\tau(u) = u\{\tau - I(u \le 0)\}$ (Koenker and Bassett, 1978). Koenker and D'Orey (1987) developed an efficient algorithm to optimize $L_\tau(\beta)$, which is available in the R package *quantreg*.

Using a similar argument given in Chamberlain (1994) for the case of independent observations, the asymptotic distribution of $N^{1/2}(\hat{\beta}_{\tau I} - \beta_\tau)$ is a normal distribution $N(0, A_\tau^{-1}\text{var}\{U_\tau(\beta_\tau)\}A_\tau^{-1})$ as $N \to \infty$, where A_τ is the expected value of the derivative of $U_\tau(\beta_\tau)$ with respect to β_τ. It is difficult to estimate the covariance matrix because A_τ may involve the unknown density functions.

Resampling method can be used to approximate the distribution of without involving any complicated and subjective nonparametric functional estimation. Let

$$\tilde{L}_\tau(\beta_\tau) = \sum_{i=1}^{N} z_i \sum_{k=1}^{n_i} \rho_\tau(y_{ik} - X_{ik}^{\mathrm{T}}\beta_\tau), \tag{6.13}$$

where $\{z_i\}_{i=1}^{N}$ is a random sample from the standard normal population. It is straightforward to show that the unconditional distribution of $N^{1/2}(\hat{\beta}_{\tau I} - \beta_\tau)$ can be approximated by the conditional distribution of $N^{1/2}(\tilde{\beta}_\tau - \beta_\tau)$ given the data (y_{ik}, X_{ik}), where $\tilde{\beta}_\tau$ is the estimator obtained from (6.13). The covariance matrix of $\hat{\beta}$ can be estimated by the empirical covariance matrix of $\tilde{\beta}_\tau$, for example, $\{\tilde{\beta}_{\tau i}, i = 1, \ldots, m\}$ is a sequence estimates of β_τ, and the empirical covariance matrix $\sum_{i=1}^{m}(\tilde{\beta}_{\tau i} - \bar{\tilde{\beta}}_\tau)(\tilde{\beta}_{\tau i} - \bar{\tilde{\beta}}_\tau)^{\mathrm{T}}/m$, where $\bar{\tilde{\beta}}_\tau$ is the sample mean of $\{\tilde{\beta}_{\tau i}, i = 1, \ldots, m\}$.

The estimating functions $U_\tau(\beta)$ are based on the independence working model assumption, hence the efficiency of $\hat{\beta}_{\tau I}$ derived from $U_\tau(\beta)$ can be enhanced if the within correlations are incorporated.

6.3.2 A Weighted Method Based on GEE

In this section, we will utilize Gaussian copulas to characterize the within correlation (Fu and Wang, 2016), which can describe various correlation structures, and construct estimating functions via the GEE method.

6.3.3 Modeling Correlation Matrix via Gaussian Copulas

It is difficult to specify and model the underlying correlation structure of $s_{i1}, \ldots, s_{in_i}$ for $i = 1, \ldots, N$. We model the correlation via Gaussian copulas. Assume that $F_i(\cdot)$

is the joint cumulative distribution function of $\epsilon_i = (\epsilon_{i1},\ldots,\epsilon_{in_i})^{\mathrm{T}}$ with marginal distributions $F_{i1}(\cdot),\ldots,F_{in_i}(\cdot)$ for $i = 1,\ldots,N$. Therefore, $U_k = F_{ik}(\epsilon_{ik})$ are uniformly distributed random variables on $[0,1]$. For given $u_1,\ldots,u_{n_i}$, we have

$$
\begin{aligned}
F_i(u_1,\cdots,u_{n_i}) &= P(\epsilon_{i1} \le u_1, \cdots, \epsilon_{in_i} \le u_{n_i}) \\
&= P(F_{i1}(\epsilon_{i1}) \le F_{i1}(u_1), \cdots, F_{in_i}(\epsilon_{in_i}) \le F_{in_i}(u_{n_i})) \\
&= P(U_1 \le F_{i1}(u_1), \cdots, U_{n_i} \le F_{in_i}(u_{n_i})) \\
&\stackrel{\text{def}}{=} C(F_{i1}(u_1), \cdots, F_{in_i}(u_{n_i})),
\end{aligned}
$$

where C is a copula, a multivariate distribution on $[0,1]^{n_i}$ with standard uniform marginal functions. According to Sklar's Theorem (Sklar, 1959), if $F_{i1}(\cdot), \cdots, F_{in_i}(\cdot)$ are absolutely continuous, C is uniquely determined on $[0,1]^{n_i}$ by $F_i(\cdot)$. Furthermore, let $F_{ik}^{-1}(\cdot)$ be the inverse function of $F_{ik}(\cdot)$, and then the copula function can be rewritten as $C(u_1,\cdots,u_{n_i}) = F_i(F_{i1}^{-1}(u_1),\ldots,F_{in_i}^{-1}(u_{n_i}))$.

Let ρ_{kl} be the correlation coefficient of s_{ik} and s_{il}, and C be a Gaussian copula. Then we have

$$
\begin{aligned}
\rho_{kl} &= \frac{1}{(\tau - \tau^2)}\left[P(\epsilon_{ik} \le 0, \epsilon_{il} \le 0) - \tau^2\right] \\
&= \frac{1}{(\tau - \tau^2)}\left[P(F_{ik}(\epsilon_{ik}) \le F_{ik}(0), F_{il}(\epsilon_{il}) \le F_{il}(0)) - \tau^2\right] \\
&= \frac{1}{(\tau - \tau^2)}\left[C(F_{ik}(0), F_{il}(0)) - \tau^2\right] \\
&= \frac{C_{kl}(\tau,\tau) - \tau^2}{(\tau - \tau^2)} = \frac{\Phi_2(\Phi^{-1}(\tau), \Phi^{-1}(\tau); \rho_{kl}) - \tau^2}{(\tau - \tau^2)}.
\end{aligned}
$$

where $\Phi_2(\cdot,\cdot;\gamma_{kl})$ denotes the standardized bivariate normal distribution with correlation coefficient $\gamma_{kl} = corr(\epsilon_{ik}, \epsilon_{il})$. Specifically, when $\gamma_{kl} = 0$, then ϵ_{ik} and ϵ_{il} are independent, and hence we have $C_{kl}(\tau,\tau) = \tau^2$ and $\rho_{kl} = 0$, which indicates that s_{ik} and s_{il} are uncorrelated. When $\tau = 0.5$, then $C_{kl}(\tau,\tau) = 1/4 + (2\pi)^{-1}\arcsin(\gamma_{kl})$, and hence $\rho_{kl} = 2/\pi \arcsin(\gamma_{kl})$.

To specify the correlation matrix R_i of $(s_{i1},\ldots,s_{in_i})$ for $i = 1,\ldots,N$, we need to specify the correlation structure of $\epsilon_i = (\epsilon_{i1},\ldots,\epsilon_{in_i})^{\mathrm{T}}$. When the correlation structure of ϵ_i is exchangeable, that is $p(\epsilon_{ik} \le 0, \epsilon_{il} \le 0)$ is a constant δ, for any $k \ne l$, the correlation coefficient of s_{ik} and s_{il} equals $\rho = (\delta - \tau^2)/(\tau - \tau^2)$, and hence the correlation matrix of $S_{\tau i} = (s_{i1},\ldots,s_{in_i})^{\mathrm{T}}$ is $R_{\tau i} = (1-\rho)I_{n_i} + \rho J_{n_i}$, where I_{n_i} is the $n_i \times n_i$ identity matrix, and J_{n_i} is an $n_i \times n_i$ matrix of 1s. Therefore, the correlation structure of $S_{\tau i}$ is exchangeable. Similarly, when the correlation structure of ϵ_i is MA(1), R_i is MA(1). When the correlation structure of ϵ_i is AR(1) or toeplitz matrix, then R_i is a toeplitz matrix. Therefore, we can construct various correlation structures for R_i via Gaussian copulas.

6.3.3.1 Constructing Estimating Functions

Let $X_i = (X_{i1},\ldots,X_{in_i})^{\mathrm{T}}$. To incorporate the within correlation and variation of the number of repeated measurements for each subject, Fu and Wang (2012) and

Fu et al. (2015) considered the following weighted estimating functions via the GEE method:

$$U_{G\tau}(\beta_\tau) = \sum_{i=1}^{N} X_i^{T} V_{\tau i}^{-1} S_{\tau i} = \sum_{i=1}^{N} X_i^{T} A_i^{-1/2} R_i^{-1} A_i^{-1/2} S_{\tau i},$$

where $A_i = \text{diag}\{\tau - \tau^2, \cdots, \tau - \tau^2\}$ is a diagonal matrix, and R_i is a working correlation matrix. When R_i is an exchangeable matrix, the inverse matrix can be written as

$$R_i^{-1} = \frac{1}{(1-\rho)}\left[I_{n_i} - \frac{\rho 1_{n_i} 1_{n_i}^{T}}{1 + (n_i - 1)\rho}\right]$$

If there is no correlation, $\rho = 0$ and $R_i^{-1} = I_{n_i}$. In this case, $U_{G\tau}(\beta_\tau)$ is equivalent to $U_\tau(\beta)$.

Suppose that $\hat{\beta}_{G\tau}$ is the resulting estimator from $U_{G\tau}(\beta_\tau)$. Under some regularity conditions, we can prove that $N^{-1/2} U_\tau(\beta_\tau) \to N(0, V_U)$, where

$$V_U = \lim_{N \to \infty} N^{-1} \sum_{i=1}^{N} X_i^{T} V_i^{-1} \text{Cov}(S_i) V_i^{-1} X_i.$$

Furthermore, $\hat{\beta}_{G\tau}$ is a consistent estimator of β_τ for a given τ, and $\sqrt{N}(\hat{\beta}_{G\tau} - \beta_\tau) \to N(0, \Sigma_{G\tau})$, where

$$\Sigma_{G\tau} = \lim_{N \to \infty} N D_\tau^{-1}(\beta)\left(\sum_{i=1}^{N} X_i^{T} V_{\tau i}^{-1} \text{Cov}(S_{\tau i}) V_{\tau i}^{-1} X_i\right)\{D_\tau^{-1}(\beta)\}^{T},$$

where $D_\tau(\beta) = \sum_{i=1}^{N} X_i^{T} V_{\tau i}^{-1} \Lambda_i X_i$, and Λ_i is an $n_i \times n_i$ diagonal matrix with the k-th diagonal element $f_{ik}(0)$. The covariance matrix $\text{Cov}(S_{\tau i})$ is unknown and can be estimated empirically. These asymptotic properties can be derived according to Jung (1996) and Yin and Cai (2005).

6.3.3.2 *Parameter and Covariance Matrix Estimation*

It is difficult to estimate the covariance matrix of parameter estimators in a quantile regression because it involves the unknown error density functions. The resampling method can be used to estimate the covariance matrix which has been described in subsection 6.3.1. However, this method often adds computation burdens.

The induced smoothing method proposed by Brown and Wang (2005) can be extended to the quantile regression (Wang et al., 2009, Fu and Wang, 2012), which can estimate parameters and their covariance simultaneously and bypass estimating the distribution function. Assume $Z \sim N(0, I_p)$, and approximate $\hat{\beta}_{G\tau}$ by $\beta_\tau + \Gamma^{1/2} Z$, where $\Gamma = O(1/N)$ is a positive definite matrix. The smoothed estimating functions of $U_{G\tau}(\beta_\tau)$ can be given as $\tilde{U}_{G\tau}(\beta_\tau) = \mathbf{E}_Z[U_{G\tau}(\beta_\tau + \Gamma^{1/2} Z)]$, where expectation is over

Z. The smoothed estimating function is:

$$\tilde{U}_{G\tau}(\beta_\tau) = \mathbf{E}_Z[U_{G\tau}(\beta_\tau + \Gamma^{1/2}Z)] = \sum_{i=1}^{N} X_i^{\mathrm{T}} V_i^{-1} \begin{pmatrix} \tau - 1 + \Phi\left(\dfrac{y_{i1} - X_{i1}^{\mathrm{T}}\beta_\tau}{\sqrt{X_{i1}^{\mathrm{T}}\Gamma X_{i1}}}\right) \\ \vdots \\ \tau - 1 + \Phi\left(\dfrac{y_{in_i} - X_{in_i}^{\mathrm{T}}\beta_\tau}{\sqrt{X_{in_i}^{\mathrm{T}}\Gamma X_{in_i}}}\right) \end{pmatrix},$$

where $\Phi(\cdot)$ is the standard normal cumulative distributed function. Because $\tilde{U}_{G\tau}(\beta_\tau)$ are smoothing functions of β_τ, we can calculate $\partial\tilde{U}_{G\tau}(\beta_\tau)/\partial\beta_\tau$ that can be used as an approximation of D_τ. Let

$$\tilde{D}_\tau(\beta_\tau) = \frac{\partial\tilde{U}_{G\tau}(\beta_\tau)}{\partial\beta_\tau} = -\sum_{i=1}^{N} X_i^{\mathrm{T}} V_i^{-1} \tilde{\Lambda}_i X_i,$$

where $\tilde{\Lambda}_i$ is an $n_i \times n_i$ diagonal matrix with the kth diagonal element $\sigma_{ik}^{-1}\phi((y_{in_i} - X_{in_i}^{\mathrm{T}}\beta_\tau)/\sigma_{ik})$, where $\phi(\cdot)$ is the standard normal density function, and $\sigma_{ik} = \sqrt{X_{ik}^{\mathrm{T}}\Gamma X_{ik}}$.

In general, the resulting estimator $\tilde{\beta}_{G\tau}$ from $\tilde{U}_\tau(\beta_\tau)$ and its covariance matrix can be obtained by iteration. Taking the exchangeable correlation working structure as an example, we give the explicit stepwise procedures for the algorithm:

Step 1: Let the estimator obtained from equation (6.12) be an initial estimate, that is $\tilde{\beta}_\tau^{(0)} = \hat{\beta}_{\tau I}$, and $\Gamma^{(0)} = I_p/N$.

Step 2: Given $\tilde{\beta}_\tau^{(k-1)}$ and $\Gamma^{(k-1)}$ from the $k-1$ step, and let $\hat{\epsilon}_{il} = y_{il} - X_{il}^{\mathrm{T}}\hat{\beta}_\tau^{(k-1)}$. Obtain $\hat{\delta}_\tau^{(k-1)}$ by

$$\hat{\delta}^{(k-1)} = \frac{\sum_{i=1}^{N}\sum_{k=1}^{n_i}\sum_{l\neq k}^{n_i} I(\hat{\epsilon}_{ik} \leq 0, \hat{\epsilon}_{il} \leq 0)}{\sum_{i=1}^{N} n_i(n_i - 1)}.$$

Step 3: Update $\tilde{\beta}_\tau^{(k)}$ and $\Gamma^{(k)}$ by

$$\tilde{\beta}_\tau^{(k)} = \tilde{\beta}_\tau^{(k-1)} + \{-\tilde{D}_\tau(\hat{\delta}^{(k-1)}, \tilde{\beta}_\tau^{(k-1)}, \Gamma^{(k-1)})\}^{-1}\tilde{U}_{G\tau}(\hat{\delta}^{(k-1)}, \tilde{\beta}_\tau^{(k-1)}, \Gamma^{(k-1)}),$$

$$\Gamma^{(k)} = \tilde{D}_\tau^{-1}(\hat{\delta}^{(k-1)}, \tilde{\beta}_\tau^{(k)}, \Gamma^{(k-1)})V_U(\hat{\delta}^{(k-1)}, \tilde{\beta}_\tau^{(k)})\tilde{D}_\tau^{-1}(\hat{\delta}^{(k-1)}, \tilde{\beta}_\tau^{(k)}, \Gamma^{(k-1)}).$$

Step 4: Repeat the above iteration Steps 2-3 until the algorithm converges.

The finial values of $\tilde{\beta}_\tau$ and Γ can be taken as the smoothed estimators of $\hat{\beta}_{G\tau}$ and its covariance matrix, respectively. Under some regularity conditions, $N^{-1/2}\{\tilde{U}_{G\tau}(\beta_\tau) - U_{G\tau}(\beta_\tau)\} = o_p(1)$, and the smoothing estimator $\tilde{\beta}_{G\tau} \to \beta_0$ in probability, and $\sqrt{N}(\tilde{\beta}_{G\tau} - \beta_\tau)$ converges in distribution to $N(0, \Sigma_{G\tau})$. Therefore, the smoothed and unsmoothed estimating functions are asymptotically equivalent uniformly in β, and the limiting distributions of the smoothed estimators coincide with the unsmoothed estimators. The induced smoothing method can be also used to the rank-based methods introduced in Section 2. More details can be found in Fu et al. (2010).

6.3.4   Working Correlation Structure Selection

Suppose that there are J working correlation matrix candidates via Gaussian copulas: R_i^j, $j = 1, \ldots, J$. Parameter estimates of β_τ can be obtained by solving J estimating equations:

$$\sum_{i=1}^{N} X_i^T A_i^{-1/2} [R_i^j]^{-1} A_i^{-1/2} S_i(\beta) = 0, \quad j = 1, \ldots, J. \tag{6.14}$$

Appropriate specification of R_i can improve the estimation efficiency of β_τ. Wang and Carey (2011) proposed a Gaussian pseudolikelihood criterion to select a working covariance model for marginal mean regression models. This criterion can be extended to select a working correlation structure in marginal quantile regression.

Let $\dim(\hat{\theta})$ be the effective dimension of θ but excluding the $\hat{\beta}$ components being zero. Substitute the pseudolikelihood Gaussian likelihood for parametric likelihood in AIC and BIC and obtain versions of AIC and BIC,

$$\text{GAIC} = -2l(\hat{\theta}) + 2\dim(\hat{\theta}),$$

$$\text{GBIC} = -2l(\hat{\theta}) + \log(m)\dim(\hat{\theta}),$$

where $l(\theta)$ is pseudolikelihood Gaussian likelihood for S_i with unknown parameters θ including β_τ and correlation parameters ρ in R_i, and is given as

$$l(\theta) = -\frac{1}{2} \sum_{i=1}^{N} \left[\log|2\pi V_i| + S_i^T V_i^{-1} S_i \right].$$

For a given set of correlation structures, we obtain estimates of β_τ and ρ and then choose the final correlation structure corresponding to the minimum value of GAIC or GBIC.

There are $J \times p$ equations in (6.14) based on J different working correlation matrices. Therefore, the number of equations is larger than the number of parameters. We can use the quadratic inference function (QIF) method proposed by Qu et al. (2000) and the empirical likelihood method to combine these equations. More details can be found in Leng and Zhang (2014) and Fu and Wang (2016).

6.3.5   Analysis of Dental Data

In this subsection, we will illustrate the method by a set of growth dental data for 11 girls and 16 boys. For each child, the distance (millimeter) from the center of the pituitary gland to the pteryomaxillary fissure was measured every two years at ages 8, 10, 12, and 14. What of interest is the relation between the recorded distance and age. Figures 6.4–6.5 indicate that the distance is linearly related to age. Furthermore, the boys' distances are larger than girls' at the same age.

We use quantile regression to model the dental data at $\tau = 0.25, 0.5, 0.75$, and $\tau = 0.95$,

$$Q_\tau(y_{ik}) = \beta_0 + \beta_1 \text{Gender}_i + \beta_2 \text{Age}_{ik} + \beta_3 \text{Gender}_i * \text{Age}_{ik},$$

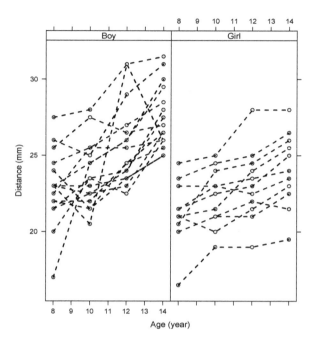

Figure 6.4 *The distance versus the measurement time for boys and girls.*

where y_{ik} and Age_{ik} are the distance and age for the i-th subject at time k, respectively, and Gender_i takes -1 for girls and 1 for boys. Parameters β_0 and β_1 denote the intercept and slope of the average growth curve for the entire group. Parameters β_1 and β_3 denote the deviations from this average intercept and slope for the group of girls and the group of boys, respectively. Therefore, the intercept and slope of the average growth curve for girls are indicated by $\beta_0 - \beta_1$ and $\beta_2 - \beta_3$, respectively. Equivalently, the intercept and slope of the average growth curve for boys are given by $\beta_0 + \beta_1$ and $\beta_2 + \beta_3$, respectively.

The parameter estimates and their standard errors are presented in Table 6.2. The results indicate that β_2 is significant at three different quantiles, which indicates that there is a linear relationship between the distance and age. The parameter β_1 is not significant, which indicates that, on average, neither girls nor boys differ significantly concerning their initial dental distance. The parameter estimate of β_3 is positive and significant at $\tau = 0.25$ but not significant at $\tau = 0.5$, 0.75, and 0.95, which indicates that boys with low distances (bottom 25%) increase faster over time than girls at $\tau = 0.25$, and there is no significant difference between boys' and girls' distance increment at $\tau = 0.5$ and 0.75. The proposed criteria have the lowest values when correlation structure is exchangeable at $\tau = 0.25$, 0.75, and 0.95. However, the GAIC and GBIC criteria reach the minimum values when correlation structure is AR(1) at $\tau = 0.5$. This indicates that kids at top or bottom 25% have more steady correlations

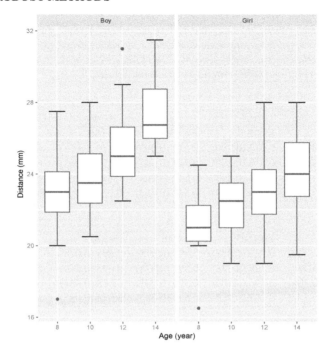

Figure 6.5 *Boxplots of distances for girls and boys.*

over time (hence exchangeable is appropriate) but correlations at medium decay fast over the years which makes the AR(1) model appropriate. This plausible explanations need to be tested on a much large dataset.

6.4 Other Robust Methods

The robust methods considered in the first two sections are based on linear models. Next we will introduce two robust methods for analyzing generalized linear models in longitudinal data. We consider the marginal model of y_{ik}, which satisfies the first two marginal moment conditions: $\mu_{ij} = g(X_{ij}^{T}\beta)$ and $\sigma_{ij}^{2} = \phi v(\mu_{ij})$, where $g(\cdot)$ and $v(\cdot)$ are specified functions, and ϕ is a scale parameter.

Fan et al. (2012) and Lv et al. (2015) proposed the following robust generalized estimating equations:

$$U_R(\beta) = \sum_{i=1}^{N} D_i^{T} M_i^{-1} h_i(\mu_i(\beta)) = 0, \qquad (6.15)$$

where $D_i = \partial\mu_i/\partial\beta$ is an $n_i \times p$ matrix, $M_i = R_i(\alpha)A_i^{1/2}$, in which $R_i(\alpha)$ is a working correlation matrix and $A_i = \phi\text{diag}(v(\mu_{i1}), \cdots, v(\mu_{in_i}))$, and $h_i(\mu_i(\beta)) = W_i[\psi(\mu_i(\beta)) - C_i(\mu_i(\beta))]$ with $\psi(\mu_i(\beta)) = \psi(A_i^{-1/2}(Y_i - \mu_i))$ and $C_i(\mu_i(\beta)) = \mathbf{E}(\psi(\mu_i(\beta)))$, which is used

Table 6.2 *Parameter estimates (Est) and their standard errors (SE) of estimators $\hat{\beta}_I$, $\hat{\beta}_{EX}$, $\hat{\beta}_{AR}$ and $\hat{\beta}_{MA}$ obtained from estimating equations with independence, exchangeable, AR(1), and MA(1) correlation structures, respectively, and frequencies of the correlation structure identified using GAIC and GBIC criteria for the dental dataset.*

$\tau = 0.25$

	β_0	β_1	β_2	β_3	GAIC	GBIC
	Est (Sd)	Est (Sd)	Est (Sd)	Est (Sd)		
$\hat{\beta}_I$	20.690 (0.299)	0.513 (0.299)	0.543 (0.063)	0.175 (0.063)	-64.790	-59.606
$\hat{\beta}_{EX}$	20.690 (0.299)	0.513 (0.299)	0.543 (0.063)	0.175 (0.063)	-69.584	-63.105
$\hat{\beta}_{AR}$	20.726 (0.298)	0.542 (0.298)	0.540 (0.063)	0.172 (0.063)	-65.797	-59.318
$\hat{\beta}_{MA}$	20.718 (0.299)	0.547 (0.299)	0.541 (0.064)	0.173 (0.064)	-65.189	-58.709

$\tau = 0.50$

	β_0	β_1	β_2	β_3	GAIC	GBIC
	Est (Sd)	Est (Sd)	Est (Sd)	Est (Sd)		
$\hat{\beta}_I$	21.877 (0.348)	0.729 (0.348)	0.591 (0.078)	0.095 (0.078)	-33.720	-28.536
$\hat{\beta}_{EX}$	21.877 (0.348)	0.729 (0.348)	0.591 (0.078)	0.095 (0.078)	-62.332	-55.852
$\hat{\beta}_{AR}$	21.918 (0.306)	0.795 (0.306)	0.595 (0.081)	0.088 (0.081)	-64.971	-58.492
$\hat{\beta}_{MA}$	21.896 (0.307)	0.803 (0.307)	0.599 (0.085)	0.078 (0.085)	-56.289	-49.810

$\tau = 0.75$

	β_0	β_1	β_2	β_3	GAIC	GBIC
	Est (Sd)	Est (Sd)	Est (Sd)	Est (Sd)		
$\hat{\beta}_I$	23.306 (0.561)	0.704 (0.561)	0.657 (0.094)	0.133 (0.094)	-72.790	-67.606
$\hat{\beta}_{EX}$	23.306 (0.561)	0.704 (0.561)	0.657 (0.094)	0.133 (0.094)	-107.511	-101.032
$\hat{\beta}_{AR}$	23.390 (0.592)	0.802 (0.592)	0.660 (0.093)	0.119 (0.093)	-101.206	-94.726
$\hat{\beta}_{MA}$	23.344 (0.593)	0.817 (0.593)	0.660 (0.094)	0.106 (0.094)	-88.079	-81.600

$\tau = 0.95$

	β_0	β_1	β_2	β_3	GAIC	GBIC
	Est (Sd)	Est (Sd)	Est (Sd)	Est (Sd)		
$\hat{\beta}_I$	25.719 (0.525)	1.335 (0.525)	0.740 (0.065)	0.068 (0.065)	-22.167	-16.984
$\hat{\beta}_{EX}$	25.719 (0.525)	1.334 (0.525)	0.740 (0.065)	0.068 (0.065)	-27.287	-20.808
$\hat{\beta}_{AR}$	25.698 (0.520)	1.327 (0.520)	0.749 (0.068)	0.070 (0.068)	-20.386	-13.907
$\hat{\beta}_{MA}$	25.719 (0.525)	1.334 (0.525)	0.740 (0.065)	0.068 (0.065)	-20.166	-13.687

to ensure the Fisher consistency of the estimate. The matrix $W_i = \text{diag}(w_{i1}, \cdots, w_{in_i})$ is a weighted diagonal matrix to downweight the effect of leverage points in the covariates.

6.4.1   Score Function and Weighted Function

The $\psi(\cdot)$ function is selected to downweight the influence of outliers in the response variable, and a common selection is robust Huber's score function $\psi_H(e) = e$ for $|e| \leq c$ and $\psi_H(e) = c\,\text{sign}(e)$ for $|e| \geq c$, where c is a tuning constant and is usually selected between 1 and 2. Assume that the cumulative distribution function of e is $F_e(\cdot)$, then we can calculate $\mathbf{E}(\psi_H(e)) = c\mathbf{E}(eI(|e| \leq c)) + c(1 - F(c) + F(-c))$. Therefore, C_i is exactly equal to zero for symmetric distributions.

Another selection of the $\psi(\cdot)$ function is the score function derived from the exponential squared loss function $\varphi(e) = 1 - \exp\{-e^2/\gamma\}$, where $\gamma > 0$ determines the degree of robustness of the estimation (Wang et al., 2013). If γ is large, we have $\varphi(e) = 1 - \exp\{-e^2/\gamma\} \approx e^2/\gamma$, hence the exponential squared loss is similar to the least squares loss function for large γ. Let the $\psi(\cdot)$ function be $e/\gamma \exp(-e^2/\gamma)$, which is based on the first derivative of $\varphi(e)$. Then the exponential squared loss function is continuous and differentiable, and hence the calculation based on the exponential squared loss function may be easier than that based on the Huber's function.

For the weighted function w_{ij}, the commonly used is Mallows-based weighted function, that is,

$$w_{ij} = \min\left\{1, \left[\frac{d}{(X_{ij} - \hat{\mu}_x)^{\mathsf{T}} S_x^{-1}(X_{ik} - \hat{\mu}_x)}\right]^{\kappa/2}\right\},$$

where d and κ are tuning constants, and $\hat{\mu}_x$ and S_x are the robust estimates of the location and covariance of x_{ik} (Rousseeuw and van Zomeren, 1990). One can use high breakdown point location and scatter estimators, such as Minimum Covariance Determinant and Minimum Volume Ellipsoid. For the tuning parameters κ and d, we can use $\kappa = 2$ and $c = \chi^2_{0.95}(p)$, which is the 0.95 quantile of a $\chi^2(p)$ distribution.

Note that if $\psi(e) = e$ and $w_{ij} = 1$ for $j = 1,\ldots,n_i, i = 1,\ldots,N$, $U_R(\beta)$ is the generalized estimating equations introduced in Chapter 4, which can be used for data without outliers.

6.4.2   Main Algorithm

The equation (6.15) must be solved by iterative methods. Hence we need to estimate the scale parameter ϕ and the correlation parameter α and then specify the initial estimate of β. If we select $\psi(e) = e$, $w_{ij} = 1$, and $R_i = I$, then (6.15) no longer depends on the scale parameter and the correlation parameter. Therefore, the solution $\hat{\beta}^{(0)}$ of (6.15) in this special case can take as an initial estimate of β. The value of $\hat{\beta}^{(0)}$ can be obtained using the functions *geese* or *gee* in the packages geese and geepack in the statistical software R.

Liang and Zeger (1986) proposed using the moment method to estimate ϕ and α in the GEE, but the moment estimates are sensitive to the outliers. We can use the following robust estimates of ϕ and α in (6.15). An robust estimate for ϕ is obtained via the median absolute deviation $\hat{\phi} = [1.483 \times \text{median}|\hat{e}_{ik} - \text{median}\{\hat{e}_{ik}\}|]^2$, where $\hat{e}_{ik} = (y_{ik} - \mu_{ik}(\hat{\beta})) / \sqrt{v(\mu_{ik}(\hat{\beta}))}$ are the Pearson residuals and $\hat{\beta}$ is the current estimate of β. For binary responses, $\phi = 1$. An robust estimate of α can be obtained by the moment method (Wang et al. 2005),

$$\hat{\alpha} = \frac{1}{NH} \sum_{i=1}^{N} \frac{1}{n_i - 1} \sum_{k \leq n_i - 1} \psi(e_{ik})\psi(e_{ik+1})$$

for the AR(1) working correlation matrix and

$$\hat{\alpha} = \frac{1}{NH} \sum_{i=1}^{N} \frac{1}{n_i(n_i - 1)} \sum_{k \neq l} \psi(e_{ik})\psi(e_{il})$$

for the exchangeable working correlation matrix, where H is the median of $\{\psi(e_{ik}), k = 1, \ldots, n_i; i = 1, \ldots, N\}$.

With an initial estimate of β, ϕ, and α, we solve (6.15) to find the estimate of β using the Fisher's scoring iterative procedure:

$$\hat{\beta}^{(k+1)} = \hat{\beta}^{(k)} - \left(\sum_{i=1}^{N} D_i^T \Sigma_i D_i \right)^{-1} \sum_{i=1}^{N} D_i^T M_i^{-1} h_i(\mu_i(\beta)) \Big|_{\beta = \hat{\beta}^{(k)}} \tag{6.16}$$

where $\Sigma_i = M_i^{-1} \Gamma_i$ and $\Gamma_i = \mathbf{E}(\partial h_i(\mu_i(\beta))/\partial \mu_i)$ for $i = 1, \ldots, N$.

Because the initial estimate is $\sqrt{N}$ consistent under the marginal model, the one-step or the final estimator from (6.16) has the same asymptotic distribution. In practice, we may simply choose to use an asymptotically equivalent estimator by stopping the iteration in a small number of steps without worrying about the convergence of (6.16). The tuning parameters c in Huber's function and γ in the exponential square loss function can be given and then fixed in the iterative procedure. An alternative way is using the data driven method to specify them, such as minimize the trace or determination of the covariance matrix of the estimators. More details can be see Wang et al. (2007) and Wang et al. (2013).

6.4.3 Choice of Tuning Parameters

The robust approach requires the specification of a loss function such as the Huber and other loss functions, and more importantly, the associated tuning parameter that determines the extent of robustness is needed. High robustness is often at the cost of efficiency loss in parameter estimation. Therefore, it makes more sense to choose a regularization (tuning) parameter according to the proportion of "outliers" in the data. Wang et al. (2007) first proposed to choose the tuning parameter in the Huber function by maximizing the asymptotical efficiency of the parameter estimation.

The data driven (or data dependent) approach leads to more efficient parameter estimation because the regularization (tuning) parameter is so chosen to reflect the proportion of outliers in the data.

More recently, this was extended to the exponential loss function and the generalized liner models for time series with heterogeneity while accounting for possible temporal correlations and irregular observations (Callens et al., 2020).

Another recent approach to determining the tuning parameters is the so called working likelihood. This term was first proposed by (Wang and Zhao, 2007) for estimating the variance parameters while the correlation structure is misspecified. The working likelihood function is deemed as a vehicle for efficient regression parameter estimation while we recognize that the data may not be generated from this likelihood function.

Treating the robust loss function as the negative log-likelihood, we can effectively obtain an appropriate regularization parameter depending on the extend of contamination in the data. Such constructed likelihood from the loss function of interest is referred to as a working likelihood. (Fu et al., 2020) derived the score functions for estimating the tuning parameters in the Huber function and bisquare functions.

Chapter 7

Clustered Data Analysis

7.1 Introduction

Suppose a population is divided into a number of groups. If any two subjects selected at random from the same group are positively correlated then each group of subjects form a cluster. Clustered data arise in varieties of practical data analytic situations, such as, epidemiology, biostatistics and medical studies. A cluster may be a village in an agricultural study, a hospital, a doctor's practice, an animal giving birth to several fetuses. For example, in a study concerned with an educational intervention program on behavior change, the data are grouped into small classes (Lui et al., 2000).

In practice-based research, multiple patients are collected per clinician or per practice. Each class or practice is a cluster. A group of genetically related members from a familial pedigree is a cluster.

7.1.1 Clustered Data

Clustered data refers to a set of measurements collected from subjects that are structured in clusters. Responses from a group of genetically related members from a familial pedigree constitute clustered data in which the responses are correlated. For example, birth weight measurements of all fetuses born to all animals in a toxicological study form a set of clustered data in which all measurements from the same animal are correlated, but measurements from two different animals are independent. Another well-known example of clustered data is small area estimation (e.g., Ran & Molina, 2015), in which each small area is considered a cluster. Clustered data is different from longitudinal data in which measurements are collected from each subject over time. Sometimes longitudinal data may be thought of as a special type of clustered data by treating a subject as a cluster in which each subject's responses over time are correlated. In longitudinal data analysis, serial correlation within a subject's responses is commonly used, while equal pairwise within cluster correlation is used in clustered data analysis. This pairwise within cluster correlation is referred to as intracluster correlation.

DOI: 10.1201/9781315153636-7

7.1.2   Intracluster Correlation

A major issue in the analysis of clustered data is that observations within a cluster are not independent, and the degree of similarity is typically measured by the intracluster correlation coefficient ρ or *ICC*. Ignoring the intraclass correlation in the analysis could lead to incorrect inference procedures, such as incorrect p-values in hypothesis testing, confidence intervals that are too small, biased estimates and effect sizes, all of which can lead to incorrect interpretation of results related intraclass correlation coefficients or effects in designed experiments. For recent reviews of intraclass correlation see Zyzanski et al. (2004) and Galbraith et al. (2010). The intraclass correlation coefficient has a long history of application in several different fields of research. In research in epidemiology, it is commonly used to measure the degree of familial resemblance with respect to biological or environmental characteristics, and in genetics it plays a central role in estimating the heritability of selected traits in animal and plant populations. In family studies it is frequently used to measure the degree of intra-family resemblance with respect to characteristics such as blood pressure, weight and height. In psychology, it plays a fundamental role in reliability theory, where observations may be collected on one or more sets of judges or assessors. Another area of application is in sensitivity analysis, where it may be used to measure the effectiveness of an experimental treatment (see Schumann and Bradley, 1957 and Donner, 1986).

Note, this is not ordinary correlation coefficient between two random variables. It is the correlation between pairs of values within the same family. How do we calculate the intraclass correlation coefficient? We illustrate in the context of familial correlation in what follows.

Suppose we want to find correlation between heights of brothers of N families in a village in which each family has n_i brothers, $i = 1, \ldots, N$. Let the heights of brothers in the N families be $x_{11}, \ldots, x_{1n_1}; \ldots; x_{N1}, \ldots, x_{Nn_N}$. The interclass correlation coefficient ρ is defined as the ordinary correlation coefficient between any two observations x_{ij} and x_{il} in the same group. As there are $M = \sum_{j=1}^{N} n_i(n_i - 1)$ pairs of values in the population, the population mean μ and variance σ^2 are defined as

$$\mu = \frac{1}{M} \sum_{i=1}^{N} (n_i - 1) \sum_{j=1}^{n_i} x_{ij}$$

and

$$\sigma^2 = \frac{1}{M} \sum_{i=1}^{N} (n_i - 1) \sum_{j=1}^{n_i} (x_{ij} - \mu)^2.$$

The covariance between pairs of values is defined as

$$v_{11} = \mathbf{E}(X_{ij} - 1)(X_{il} - 1) = \frac{1}{M} \sum_{i=1}^{N} \sum_{j \neq l=1}^{n_i} (x_{ij} - \mu)(x_{il} - \mu).$$

This, after simplification, can be written as

$$\upsilon_{11} = \frac{1}{M}\left[\sum_{i=1}^{N} n_i^2(\mu_i - \mu)^2 - \sum_{i=1}^{N}\sum_{j=1}^{n_i}(x_{ij}-\mu)^2\right],$$

where $\mu_i = \frac{1}{n_i}\sum_{j=1}^{n_i} x_{ij}$. The intraclass correlation coefficient is defined as

$$\rho = \frac{E(X_{ij}-1)(X_{il}-1)}{\sqrt{Var(X_{ij})Var(X_{il})}} = \frac{\upsilon_{11}}{\sigma^2} = \frac{\sum_{i=1}^{N} n_i^2(\mu_i-\mu)^2 - \sum_i\sum_j(x_{ij}-\mu)^2}{\sum_{i=1}^{N}(n_i-1)\sum_{j=1}^{n_i}(x_{ij}-\mu)^2}.$$

Note, υ_{11} is the covariance between any two observations within the same cluster for all clusters, $Var(X_{ij}) = Var(X_{il}) = \sigma^2$.

Example 1. As an example, consider the data in Kendall and Stuart (1979, p322). Data on heights in inches of three brothers in five families are 74, 70, 72; 70, 71, 72; 71, 72, 72; 68, 70, 70; 71, 72, 70.

For these data $M = 30$, $n_1 = n_2 = n_3 = n_4 = n_5 = 3$, $\sum_{j=1}^{3} x_{1j} = 211$, $\sum_{j=1}^{3} x_{2j} = 213$, $\sum_{j=1}^{3} x_{3j} = 215$, $\sum_{j=1}^{3} x_{34j} = 208$, $\sum_{j=1}^{3} x_{5j} = 216$, $\mu_1 = 70.33$, $\mu_2 = 71$, $\mu_3 = 71.66$, $\mu_4 = 69.33$, and $\mu_5 = 72$. Using these we obtain

$$\mu = \frac{1}{M}\sum_{i=1}^{N}(n_i-1)\sum_{i=1}^{n_i} x_{ij} = 70.86, \quad \sum_{j=1}^{3}(x_{1j}-\mu)^2 = 5.4988,$$

$$\sum_{j=1}^{3}(x_{2j}-\mu)^2 = 2.0588, \quad \sum_{j=1}^{3}(x_{3j}-\mu)^2 = 2.6188,$$

$$\sum_{j=1}^{3}(x_{4j}-\mu)^2 = 9.6588, \quad \sum_{j=1}^{3}(x_{5j}-\mu)^2 = 5.8988.$$

Then,

$$\sigma^2 = \frac{1}{M}\sum_{i=1}^{N}(n_i-1)\sum_{j=1}^{n_i}(x_{ij}-\mu)^2 = \frac{2*25.734}{30} = 1.7156$$

and

$$\rho = \frac{\sum_{i=1}^{N} n_i^2(\mu_i-\mu)^2 - \sum_{i=1}^{N}\sum_{j=1}^{n_i}(x_{ij}-\mu)^2}{\sum_{i=1}^{N}(n_i-1)\sum_{j=1}^{n_i}(x_{ij}-\mu)^2} = \frac{41.229-25.734}{2*25.724} = 0.301.$$

7.2 Analysis of Clustered Data: Continuous Responses

7.2.1 Inference for Intraclass Correlation from One-way Analysis of Variance

Inference procedures for ρ, as first noted by Fisher (1925), are closely related to the more general problem of inference procedures for variance components. Variance components estimation is not the subject in this chapter, so we will briefly mention what are variance components in the context of estimating the interclass correlation.

For details regarding variance components we refer the reader to Sahai (1979), Shrout and Fleiss (1979), Khuri and Sahai (1985), Sahai et al. (1985), Griffin and Gonzalez (1995), and Searle et al. (2009).

Suppose that we have data on L groups or families, the ith family having n_i observations. Let y_{ij} be the observation on the jth member of the ith family. Then, y_{ij} can be represented by a random effects model

$$y_{ij} = \mu + a_i + e_{ij}, \tag{7.1}$$

where μ is the grand mean of all the observations in the population, $\{a_i\}$ is the random effect of the ith family, and $\{e_{ij}\}$ is the error in observing y_{ij}. The random effects $\{a_i\}$ are identically distributed with mean 0 and variance σ_a^2, the residual errors e_{ij} are identically distributed with mean 0 and variance σ_e^2, and the $\{a_i\}$, $\{e_{ij}\}$ are completely independent. The variance of y_{ij} is then given by

$$\sigma^2 = \sigma_a^2 + \sigma_e^2,$$

and the covariance between y_{ij} and y_{il} $(l \ne j)$ is

$$\text{Cov}(\mu + a_i + e_{ij}, \mu + a_i + e_{il}) = \text{Var}(a_i) = \sigma_a^2.$$

The intraclass correlation ρ is defined as

$$\rho = \sigma_a^2/(\sigma_a^2 + \sigma_e^2) = \sigma_a^2/\sigma^2.$$

This implies that $Cov(y_{ij}, y_{il}) = \sigma^2 \rho$. The quantities σ_a^2 and σ_e^2 are called the variance components. The above random effects model can simply be written as

$$y_i = (y_{i1}, y_{i2}, \ldots, y_{in_i})^{\text{T}} \sim N(\nu_i, \Sigma_i), \tag{7.2}$$

where $N(,)$ stands for a multivariate normal density, $\nu_i = (\mu, \ldots, \mu)^{\text{T}}$ is a vector of length n_i, and
$$\Sigma_i = \sigma^2 \{(1-\rho)I_i + \rho J_i\}$$

is a $n_i \times n_i$ matrix, I_i denotes a $n_i \times n_i$ identity matrix, J_i a $n_i \times n_i$ matrix containing only ones.

Let $n_0 = (N^* - \sum_{i=1}^{N} n_i^2/N^*)/(L-1)$, where $N^* = \sum_{i=1}^{N} n_i$ is the total number of observations and p is the number of families studied, $SSB = \sum_{i=1}^{N} n_i(y_{ij} - \bar{y}_{i.})^2$, and $SSW = \sum_{i=1}^{N} \sum_{j=1}^{n_i} (y_{ij} - \bar{y}_{i.})^2$. Then, the following table shows the analysis of variance corresponding to this model (see Donner, 1986)

Source	Degrees of freedom	Sum of squares	Mean square	E(MS)
Between families	$N-1$	SSB	MSB	$n_0\sigma_a^2 + \sigma_e^2$
Within families	$N^* - N$	SSW	MSW	σ_e^2

Now, an unbiased estimate of σ_e^2 is MSW and that of $n_0\sigma_a^2 + \sigma_e^2$ is MSB. Solving these two equations, an estimate of σ_a^2 is $\hat{\sigma}_a^2 = (\text{MSB} - \text{MSW})/n_0$ and that of ρ is

$$\hat{\rho} = \frac{\text{MSB} - \text{MSW}}{\text{MSB} + (n_0 - 1)\text{MSW}}.$$

Example 2: Let $n_0 = (N^* - \sum_{i=1}^{N} n_i^2/N^*)/(L-1)$, where $N^* = \sum_{i=1}^{N} n_i$ is the total number of observations and L is the number of families studied, $\text{SSB} = \sum_{i=1}^{N} n_i(\bar{y}_{i.} - \bar{y}_{..})^2$ following standard notation, and $\text{SSW} = \sum_{i=1}^{N} \sum_{j=1}^{n_i} (y_{ij} - \bar{y}_{i.})^2$. Then, following table on page 19 of Donner and Koval (1980) we have now, an unbiased estimate of σ_e^2 is MSW and that of $n_0\sigma_a^2 + \sigma_e^2$ is MSB. Solving these two equations, an estimate of σ_a^2 is $\hat{\sigma}_a^2 = (\text{MSB} - \text{MSW})/n_0$ and that of ρ is

$$\hat{\rho} = \frac{\text{MSB} - \text{MSW}}{\text{MSB} + (n_0 - 1)\text{MSW}}.$$

Now, consider a hypothetical data set of two judges testing 8 different wines assigning scores from 0 to 9, y_{ij}, $i = 1, 2, j = 1, 2, ..., 8$ as (4,2), (1,3), (3,8), (6, 4), (6,5), (7,5), (8,7), (9,7). Find the intraclass correlation of the judges assigning the scores. For these data we obtain MSB = 0.5625, MSW = 5.776, hence

$$\hat{\rho} = \frac{\text{MSB} - \text{MSW}}{\text{MSB} + (n_0 - 1)\text{MSW}} = -0.127$$

7.2.2   Inference for Intracluster Correlation from More General Settings

Donner and Koval (1980) deal with estimation of common interclass correlation in samples of unequal size from a multivariate normal population with no covariances in the framework of a random effects linear model (7.1).

Donner and Bull (1983) considered maximum likelihood estimation of a common interclass correlation in two independent samples drawn from multivariate normal populations. In particular they considered the situation with a different number of classes per population but an equal number of observations per class within a population.

Suppose we have m_i classes of observations y_{ijk} ($k = 1, 2, ..., n_i$; $j = 1, 2, ..., m_i$; $i = 1, 2, ..., L$) with n_i observations per class in the ith population. We assume that the y_{ijk} are distributed according to a multivariate normal distribution about the same mean μ_i and same variance σ_i^2 within the ith population, in such a way that the observations, y_{ijk} and y_{ijl}, in the same class have a common correlation, ρ.

Let I_i be a $n_i \times n_i$ identity matrix, J_i be a $n_i \times n_i$ unit matrix, $\Sigma_i = \{(1-\rho)I_i + \rho J_i\}$ be a matrix of dimension $n_i \times n_i$, and $y_{ij} = (y_{ij1}, y_{ij2}, ..., y_{ijn_i})^T$. Then we may write

$$y_{ij} \sim N(\nu_i, \sigma_i^2 \Sigma_i), \tag{7.3}$$

where $N(\cdot)$ denotes the multivariate normal density, $v_i = (\mu_i, \mu_i, \ldots, \mu_i)^{\mathrm{T}}$ is a vector of length n_i. For $L = 2$ model (7.3) is the model considered by Donner and Bull (1983), otherwise it is a generalization of their model for L groups or populations.

Rosner (1984) developed methods for performing multiple regression analyses and multiple logistic regression analyses on ophthalmologic data with normally and binomially distributed outcome variables, while accounting for the interclass correlation between eyes. Here we deal with normally distributed outcome variables in which we assume a nested data structure with L primary units of analysis, where within the ith primary unit of analysis there are t_i secondary units of analysis (or subunits), $i = 1, \ldots, L$. In ophthalmologic data the primary unit of analysis is a person and the secondary units are the eyes with $t_i = 2$.

Let y_{ij} be the measure on the jth secondary unit of the ith primary unit, $j = 1, \ldots, t_i, i = 1, \ldots, L$. Then Rosner (1984) considered the multiple regression model

$$y_{ij} = \beta_0 + \sum_{k=1}^{p-1} \beta_k x_{ijk} + e_{ij},$$

where variance of e_{ij} is σ^2 and covariance between e_{ij} and e_{il} is ρ, $i = 1, \cdots, L$, $j \neq l = 1, \ldots, t_i$. This model can conveniently be written as

$$Y_i \sim N(X_i\beta, \sigma^2\Sigma_i), \tag{7.4}$$

where $Y_i = (y_{i1}, \ldots, y_{in_i})^{\mathrm{T}}$, N follows the multivariate normal distribution, X_i is a $n_i \times p$ matrix of covariates, β is a $p \times 1$ vector of regression coefficients and $\Sigma_i = \{(1-\rho)I_i + \rho J_i\}$ is a matrix of dimension $n_i \times n_i$.

Munoz, Rosner and Carey (1986) considered estimation of common interclass correlation where the purpose was to study possible common effect of several independent variables in a regression context after allowing for heterogeneous interclass correlations. Their regression model is

$$y_{ij} = X_{ij}\beta + e_{ij},$$

where y_{ij} is an $n_{ij} \times 1$ vector of values of the dependent variable of the individuals belonging to the jth family of type i, X_{ij} is a $n_{ij} \times p$ matrix of covariances, $(p-1)$ is the number of covariates, β is the $p \times 1$ vector of regression coefficients, and the vectors of residuals e_{ij} are assumed to be independent and to follow a multivariate normal distribution with mean zero and covariance matrix $\sigma^2 V_{ij}$, where V_{ij} is an $n_{ij} \times n_{ij}$ matrix with 1 on the diagonal and ρ_i everywhere else, so that the interclass correlations depend on the type of family. This model can conveniently be written as

$$y_{ij} \sim N(X_{ij}\beta, \sigma^2 V_{ij}), \tag{7.5}$$

where $Y_{ij} = (y_{ij1}, \ldots, y_{ijn_{ij}})^{\mathrm{T}}$, X_{ij} is a $n_{ij} \times p$ matrix of $(p-1)$ covariates, β is a $p \times 1$ vector of regression coefficients and $V_{ij} = \{(1-\rho_i)I_{ij} + \rho_i J_{ij}\}$ is a matrix of dimension $n_{ij} \times n_{ij}$.

Models (7.2) to (7.5) are all special cases of the generalized regression model (Paul, 1990),

$$Y_{ij} \sim N(X_{ij}\beta_i, \sigma_i^2 V_{ij}), \tag{7.6}$$

where $Y_{ij} = (Y_{ij,1}, \ldots, Y_{ijn_{ij}})^T$, X_{ij} is a $n_{ij} \times p$ matrix of $(p-1)$ covariates, β_i is a $p \times 1$ vector of regression coefficients and $V_{ij} = (1 - \rho_i)I_{ij} + \rho_i J_{ij}$ is a matrix of dimension $n_{ij} \times n_{ij}$.

Now, if we put $\sigma_i^2 = \sigma^2$, $\beta_i = \beta = (\beta_0, \beta_1, \cdots, \beta_{p-1})$ in Model (7.6), we obtain model (7.5). In addition if we let $\rho_i = \rho$, we obtain model (7.4). Further, if we put $p = 1$, $n_{ij} = n_i$, $m_i = 1$, $\rho_i = \rho, \sigma_i^2 = \sigma^2$, and $\beta_i = v_i = v$ in Model (7.6), we obtain model (7.2). In Model (7.6), if we put $p = 1$, $\rho_i = \rho$, $\beta_i = v_i$, and $n_{ij} = n_i$, we obtain model (7.3), where $v_i = (\mu_i, \mu_i, \ldots, \mu_i)^T$.

7.2.3 Maximum Likelihood Estimation of the Parameters

It is well known that maximum likelihood estimates of the parameters are obtained by taking derivatives of the log likelihood with respect to the parameters of interest, equating the resulting derivatives to zero, and finally solving these estimating equations.

It can be seen that minus twice the log-likelihood for L independent samples from Model (7.6) can be conveniently written as

$$-2l = \sum_{i=1}^{L} \sum_{j=1}^{m_i} \left[n_{ij} \log \sigma_i^2 + (n_{ij} - 1) \log(1 - \rho_i) + \log\{1 + (n_{ij} - 1)\rho_i\} \right]$$

$$+ \sum_{i=1}^{L} \sum_{j=1}^{m_i} \left[\frac{(Y_{ij} - X_{ij}\beta_i)^T I_{ij}(Y_{ij} - X_{ij}\beta_i)}{(1 - \rho_i)\sigma_i^2} - \frac{\rho_i(Y_{ij} - X_{ij}\beta_i)^T J_{ij}(Y_{ij} - X_{ij}\beta_i)}{(1 - \rho_i)\{1 + (n_{ij} - 1)\rho_i\}\sigma_i^2} \right].$$

Then, following the outlines above the maximum likelihood estimates (or more accurately, a solution to the ML equations) of β_i and σ_i^2 given ρ_i are

$$\hat{\beta}_i|\rho_i = \left(\sum_{j=1}^{m_i} X_{ij}^T W_{ij} X_{ij} \right)^{-1} \left(\sum_{j=1}^{m_i} X_{ij}^T W_{ij} Y_{ij} \right) \tag{7.7}$$

and

$$\hat{\sigma}_i^2|\rho_i = \sum_j \frac{SS_{ij} - \rho_i n_{ij} SST_{ij} t_{ij}^{-1}}{(1 - \rho_i)N_{ij}}, \tag{7.8}$$

respectively, where $N_i = \sum_j n_{ij}$, W_{ij} is the inverse of V_{ij} with $(t_{ij} - \rho_i)/\{(1 - \rho_i)t_{ij}\}$ on the diagonal and $-\rho_i/\{(1 - \rho_i)t_{ij}\}$ everywhere else, $t_{ij} = 1 + (n_{ij} - 1)\rho_i$,

$$SS_{ij} = (Y_{ij} - X_{ij}\beta_i)^T I_{ij}(Y_{ij} - X_{ij}\beta_i),$$

and

$$SST_{ij} = (Y_{ij} - X_{ij}\beta_i)^T J_{ij}(Y_{ij} - X_{ij}\beta_i)/n_{ij}.$$

Further equating $-2\partial l/\partial \rho_i$ to zero, and after some algebra the estimating equation for ρ_i, is

$$\frac{N_i^{-1}\sum_{j=1}^{m_i}[\text{SS}_{ij} - n_{ij}\text{SST}_{ij}t_{ij}^{-2}1 + (n_{ij}-1)\rho_i^2]}{\sum_{j=1}^{m_i}(SS_{ij} - \rho_i n_{ij}\text{SST}_{ij}t_{ij}^{-1})} - \rho_i \sum_{j=1}^{m_i} n_{ij}(n_{ij}-1)t_{ij}^{-1} = 0, \qquad (7.9)$$

where $\rho_i \epsilon \cap_j (-1/(n_{ij}-1), 1)$. Note that this equation involves only ρ_i as unknown, while β_i in SS_{ij} and SST_{ij} involves data and ρ_i. Thus, the maximum likelihood estimate of ρ_i is obtained by solving equation (7.9) iteratively. Equation (7.9) is the optimal estimating equation in the sense of Godambe (1960) and Bhapkar (1972). That is, as a maximum likelihood equation, the estimating equation for ρ_i is unbiased and fully efficient. Denote the estimate of ρ_i obtained by solving equation (7.9) by $\hat{\rho}_i$. Then, replacing ρ_i on the right hand side of equations (7.7) and (7.8) by $\hat{\rho}_i$, the maximum likelihood estimates of β_i and σ_i^2 are obtained.

Maximum likelihood (ML) estimates of the parameters of models (7.2)–(7.5) can be obtained as special cases of equations (7.7)–(7.9). So the ML estimates of β and σ^2 of model (7.5) are

$$\hat{\beta} = \left(\sum_{i=1}^{L}\sum_{j=1}^{m_i} X_{ij}^{\text{T}} W_{ij} X_{ij}\right)^{-1} \sum_{i=1}^{L}\sum_{j=1}^{m_i} X_{ij}^{\text{T}} W_{ij} Y_{ij} \qquad (7.10)$$

and

$$\hat{\sigma}^2 = \sum_{i=1}^{L}\sum_{j=1}^{m_i} \frac{\text{SS}_{ij} - \rho_i n_{ij}\text{SST}_{ij}t_{ij}^{-1}}{\tilde{N}(1-\rho_i)}, \qquad (7.11)$$

respectively, where $\tilde{N} = \sum_i \sum_j n_{ij}$ and the estimating equation for ρ_i is

$$\frac{\tilde{N}(1-\rho_i)^{-1}\sum_j[\text{SS}_{ij} - n_{ij}\text{SST}_{ij}t_{ij}^{-2}\{1 + (N_{ij}-1)\rho_i^2\}]}{\sum_i(1-\rho_i)^{-1}\sum_j(\text{SS}_{ij} - \rho_i n_{ij}\text{SST}_{ij}t_{ij}^{-1})} - \rho_i \sum_j n_{ij}(n_{ij}-1)t_{ij}^{-1} = 0,$$

$$(7.12)$$

where $\rho_i \in \cap_j(-1/(n_{ij}-1), 1)$, $\text{SS}_{ij} = (Y_{ij} - X_{ij}\beta_i)^{\text{T}} I_{ij}(Y_{ij} - X_{ij}\beta_i)$, $\text{SST}_{ij} = (Y_{ij} - X_{ij}\beta_i)^{\text{T}} J_{ij}(Y_{ij} - X_{ij}\beta_i)/n_{ij}$, and $t_{ij} = 1 + (n_{ij}-1)\rho_i$.

Now, let Ω_{ij} be the inverse of Σ_{ij} with $(t_{ij}-\rho)/\{(1-\rho)t_{ij}\}$ in the diagonal and $-\rho/\{(1-\rho)t_{ij}\}$ everywhere else; $t_{ij} = 1 + (n_{ij}-1)\rho$ and SS_{ij} and SST_{ij} are as defined for equations (7.10) and (7.11) above. Then the maximum likelihood estimates of β and σ^2 of model (7.4) are respectively

$$\hat{\beta} = \left(\sum_i \sum_j X_{ij}^{\text{T}} \Omega_{ij} X_{ij}\right)^{-1} \left(\sum_i \sum_j X_{ij}^{\text{T}} \Omega_{ij} Y_{ij}\right) \qquad (7.13)$$

and

$$\hat{\sigma}^2 = \sum_i \sum_j (\text{SS}_{ij} - \rho n_{ij}\text{SST}_{ij}t_{ij}^{-1})/\{N(1-\rho)\}. \qquad (7.14)$$

The estimating equation for ρ is

$$\frac{\tilde{N}\sum_i\sum_j[SS_{ij} - n_{ij}SST_{ij}t_{ij}^{-2}\{1 + (N_{ij} - 1)\rho_i^2\}]}{\sum_i\sum_j(SS_{ij} - \rho n_{ij}SST_{ij}t_{ij}^{-1})} - \rho\sum_i\sum_j n_{ij}(n_{ij} - 1)t_{ij}^{-1} = 0, \quad (7.15)$$

where $\rho_i \in \cap_j(-1/(n_{ij} - 1), 1)$.

The maximum likelihood estimates of the parameters of Model (7.2) can be obtained from equations (7.13)-(7.15) by taking $m_i = 1$, that is subscript j is not required, so that $t_{ij} = t_i = 1 + (n_i - 1)\rho$, $\rho_i = \rho$, $X_i = (1,\ldots,1)^T$ is a vector of dimension n_i and $\beta = \mu$. Thus, for given ρ the maximum likelihood estimates of μ and σ^2 are

$$\hat{\mu} = \frac{\sum_i n_i\bar{y}_i t_i^{-1}}{\sum_i n_i t_i^{-1}} \quad (7.16)$$

and

$$\hat{\sigma}^2 = \frac{SS - \rho\sum_i n_iSST_i t_i^{-1}}{N(1 - \rho)}, \quad (7.17)$$

where $SS = \sum_i\sum_k(y_{ik} - \mu)^2$ and $SST_i = n_i(\bar{y}_{i.} - \mu)^2$, and the estimating equation for ρ is

$$\frac{\tilde{N}[SS - \sum_i n_iSST_i\{1 + (n_i - 1)\rho^2\}t_i^{-2}]}{SS - \rho\sum_i n_iSST_i t_i^{-1}} - \rho\sum_i\frac{n_i(n_i - 1)}{1 + (n_i - 1)\rho} = 0, \quad (7.18)$$

where $\rho \in \cap_i(-1/(n_i - 1), 1)$.

Similarly, the ML estimates of μ_i and σ_i^2 of model (7.3) are respectively

$$\hat{\mu}_i = \sum_j\sum_k y_{ijk}/n_i m_i = \bar{y}_{i..}$$

and

$$\sigma_i^2 = \frac{(SS_i - \rho n_iSST_i t_i^{-1})}{n_i m_i(1 - \rho)},$$

where

$$SS_i = \sum_j\sum_k(y_{ijk} - \bar{y}_{i..})^2,$$

and

$$SST_i = n_i\sum_j(\bar{y}_{ij.} - \bar{y}_{i..})^2, t_i = 1 + (n_i - 1)\rho.$$

Further, after simplification, the estimating equation for ρ can be expressed as

$$\sum_i\frac{m_i n_i(n_i - 1)(\rho - r_i)}{(1 - \rho R_i)\{1 + (n_i - 1)\rho\}} = 0, \quad (7.19)$$

where

$$r_i = \frac{n_iSST_i - SS_i}{(n_i - 1)SS_i} \quad (7.20)$$

is the sample intraclass correlation from the ith population, $R_i = 1 - (n_i - 1)(1 - r_i)$, and $\rho \in \cap_i(-1/(n_i - 1), 1)$.

7.2.4 Asymptotic Variance

It can be easily found that for all the models the β parameters and the (σ^2, ρ) parameters are orthogonal, i.e.

$$-\mathbf{E}\left(\frac{\partial^2 l}{\partial \beta \partial \sigma^2}\right) = -\mathbf{E}\left(\frac{\partial^2 l}{\partial \beta \partial \rho}\right) = 0.$$

The consequence of such orthogonality is that the estimates $\hat{\beta}$ and $(\hat{\sigma}^2, \hat{\rho})$ are asymptotically independent (Cox and Reid, 1987). The asymptotic variance of $\hat{\rho}$ is thus obtained from the inverse of the Fisher information matrix of (σ^2, ρ). For Model (7.6), it can be shown that

$$-\mathbf{E}\left(\frac{\partial^2 l}{\partial \sigma_i^4}\right) = \frac{N_i}{2\sigma_i^4}, \quad -\mathbf{E}\left(\frac{\partial^2 l}{\partial \sigma_i^2 \partial \rho_i}\right) = \frac{-A_i}{2\sigma_i^2(1-\rho_i)}, \quad -\mathbf{E}\left(\frac{\partial^2 l}{\partial \rho_i^2}\right) = \frac{B_i}{2(1-\rho_i)^2},$$

where $N_j = \sum_j n_{ij}$,

$$A_i = \sum_j \frac{n_{ij}(n_{ij}-1)\rho_i}{1+(n_{ij}-1)\rho_i}$$

and

$$B_i = \sum_j \frac{n_{ij}(n_{ij}-1)\{1+(n_{ij}-1)\rho_i^2\}}{\{1+(n_{ij}-1)\rho_i\}^2}.$$

From the inverse of the Fisher information matrix of (σ_i^2, ρ) it can be seen that

$$\text{var}(\hat{\rho}_i) = \frac{2N_i(1-\rho_i)^2}{N_i B_i - A_i^2}.$$

Similarly, it can be shown that for Model (7.5)

$$\text{var}(\hat{\rho}_i) = \frac{2\tilde{N}(1-\rho_i)^2}{\tilde{N} B_i - A_i^2},$$

where $\tilde{N} = \sum_i N_i$; for Model (7.4),

$$\text{var}(\hat{\rho}) = \frac{2\tilde{N}(1-\rho)^2}{\tilde{N} B - A^2},$$

where

$$A = \sum_i \sum_j \frac{n_{ij}(n_{ij}-1)\rho}{1+(n_{ij}-1)\rho_i}$$

and

$$B = \sum_i \sum_j \frac{n_{ij}(n_{ij}-1)\{1+(n_{ij}-1)\rho^2\}}{\{1+(n_{ij}-1)\rho\}^2};$$

for Model (7.2),

$$\text{var}(\hat{\rho}) = \frac{2N(1-\rho)^2}{ND - C^2},$$

where $n = \sum_i n_i$,

$$C = \sum_i \frac{n_i(n_i - 1)\rho}{1 + (n_i - 1)\rho} \quad \text{and} \quad D = \sum_i \frac{n_i(n_i - 1)\{1 + (n_i - 1)\rho^2\}}{\{1 + (n_i - 1)\rho\}^2},$$

and finally for Model (7.3),

$$\text{var}(\hat{\rho}) = \frac{2(1-\rho)^2}{\sum_i m_i n_i(n_i - 1)}\{1 + (n_i - 1)\rho\}^{-2}.$$

7.2.5 Inference for Intracluster Correlation Coefficient

Inference for the intraclass correlation can be done by constructing confidence intervals or conducting hypothesis tests for ρ or ρ_i using the asymptotic standard errors. The estimating equations for ρ can be solved by a subroutine in R.

Example 3

Table 7.1 Artificial data of Systolic blood pressure of children of 10 families

					Families					
	1	2	3	4	5	6	7	8	9	10
Systolic	118	111	107	115	127	126	90	140	110	109
Blood	130	140	110	115	122	123	99	120	105	
Pressure		125	125		140	109			138	
		100				134			103	
						88				

The purpose of the example here is to illustrate that the estimating equation for ρ produces unique solution within the permitted range. The estimating equation was solved by using R and solution was obtained within 9 iterations.

We consider Model IV, page 550 of Paul, 1990, where $\mathbf{Y}_i = (Y_{i1}, \ldots, Y_{in_i})', \nu = (\mu, \ldots, \mu)$ and $\Sigma_i = (1 - \rho)I_i + \rho J_i$, where I_i is a $n_i \times n_i$ identity matrix and J_i is a $n_i \times n_i$ matrix of only ones.

Now, let $t_i = 1 + (n_i - 1)\rho$,

$$\mu = \frac{\sum_i n_i \bar{y}_i / t_i}{\sum_i n_i / t_i}$$

$$\sigma^2 = \frac{SS - \rho \sum_i n_i SST_i / t_i}{N(1 - \rho)}, \qquad N = \sum_i n_i$$

where,

$$SS = \sum_i \sum_k (y_{ik} - \mu)^2 \qquad SST_i = n_i(\bar{y}_i - \mu)^2$$

and the estimating equation for ρ is,

$$\frac{N\left[SS - \sum_i n_i SST_i\{1 + (n_i - 1)\rho^2\}/t_i^2\right]}{SS - \rho \sum_i n_i SST_i/t_i} - \rho \sum_i \frac{n_i(n_i - 1)}{1 + (n_i - 1)\rho} = 0$$

where,

$$\rho \in \cap_i\left(-\frac{1}{n_i - 1}, 1\right) \quad \text{and}$$

$$Var(\hat{\rho}) = \frac{2(1 - \hat{\rho})^2}{\sum_i n_i(n_i - 1)}\{1 + (n_i - 1)\rho\}^{-2}$$

The final solution is $\hat{\rho} = 0.0266$ with standard error=0.1596. The solution took 6 iterations.

7.2.6   Analysis of Clustered or Intralitter Data: Discrete Responses

Discrete data in the form of proportions arise in toxicological experiments from littermates. A litter is generally a pregnant animal and the littermates are usually the fetuses within an animal. Investigators studying the teratogenic, mutagenic or carcinogenic effects of chemical agents in laboratory animals frequently obtain data from littermates. As Paul (1982) described: Data arise in teratological experiments in which a number of pregnant female animals are assigned to different groups. These generally consist of a control group and a number of groups treated with varying doses of a compound. Each fertilized egg results in either a resorption, an early foetal death, a late foetal death or a live fetus. Treatments may affect the incidence of each of these characteristics, but the principal aim of the experiment is to determine if treatments affect the incidences of abnormalities in live fetuses as the littermates may receive an indirect exposure by the treatment of one parent with the compound under study. More details are given in a review paper by Haseman and Kupper (1979).

7.2.7   The Models

The simplest model to describe toxicological data in the form of proportions is the binomial model. However, it happens quite often that these types of data show greater variability than predicted by the simple binomial model and the reason for this variability depends on the form of study. Weil (1970) observed that if the experimental units of the data are litters of animals then "litter effect", that is, the tendency of animals in the same litter to respond more similarly than animals from different litters contribute to greater variability than predicted by the simple model. This litter effect is known as intra-litter correlation coefficient. Several models, such as the beta-binomial model (BB) (Williams, 1975; Crowder, 1978), the additive binomial and the multiplicative binomial (MB) models (Altham, 1978) and the correlated binomial model (CB) (Kupper and Haseman, 1978) have been proposed in the literature to account for this intra-litter correlation coefficient. The additive binomial and the correlated binomial models are identical (Paul, 1982). However, the beta-binomial model is the most popular and widely used as it often fits the toxicological proportion data much better than the other models (Paul, 1982).

Consider a toxicological experiment consisting of m animals or litters, treated by a compound (treatment), the ith litter giving birth to n_i fetuses of which y_i fetuses are affected (for example, malformed). Thus, the observed proportion of fetuses affected by the treatment for the ith litter is y_i/n_i, $i = 1, \ldots, m$. Let p be the overall proportion of fetuses affected by the treatment. We assume that $y_i|p \sim binomial(n_i, p)$ and p is a random variable having a beta distribution with probability density function

$$f(p) = \frac{1}{B(\alpha,\beta)} p^{\alpha-1}(1-p)^{\beta-1}.$$

Then, unconditionally, y_i has a beta-binomial distribution with probability mass function

$$Pr(y_i) = \binom{n_i}{y_i} \frac{B(y_i + \alpha, n_i + \beta - y_i)}{B(\alpha,\beta)}. \tag{7.21}$$

The mean and variance of y_i are $n_i\left(\frac{\alpha}{\alpha+\beta}\right)$ and $\frac{n_i\alpha\beta(\alpha+\beta+n_i)}{(\alpha+\beta)^2(\alpha+\beta+1)}$, respectively. Now, define $\pi = \frac{\alpha}{\alpha+\beta}$, $\omega = \frac{1}{\alpha+\beta}$ and $\phi = \frac{\omega}{1+\omega}$. Then the mean and the variance of y_i can be expressed as $n_i\pi$ and $n_i\pi(1-\pi)[1+(n_i-1)\phi]$. The parameter ϕ is the extra-dispersion parameter which is also known as the intra-litter correlation parameter. In practical contexts the interest is to estimate the parameters π and ϕ. The beta-binomial distribution assumes that the intra-litter correlation parameter ϕ is positive. However, Prentice (1986) argued that ϕ can also assume negative values and proposed the extended beta-binomial distribution where ϕ can assume positive as well as negative values with the range given below. Using the above parametrization, further simplification, ϕ can assume positive as well as negative values. The probability mass function of the extended beta-binomial distribution is

$$Pr(y_i|\pi, \phi) = \binom{n_i}{y_i} \frac{\prod_{r=0}^{y_i-1}[\pi(1-\phi)+r\phi] \prod_{r=0}^{n_i-y_i-1}[(1-\pi)(1-\phi)+r\phi]}{\prod_{r=0}^{n_i-1}[(1-\phi)+r\phi]} \tag{7.22}$$

with $0 \le \pi_i \le 1$, and $\phi_i \ge \max\left[-\pi_i/(n_{ij}-1), -(1-\pi_i)/(n_{ij}-1)\right]$. Then ϕ can take positive as well as negative values, and can be justified from the fact that fetuses for the same litter can compete for the same resources within the mother's womb.

7.2.8   Estimation

A number of methods for the estimation of π and ϕ are available. Here we give two of the most popular methods, namely, the method of moments and the maximum likelihood estimates. For a comprehensive description of the methods available see Paul and Islam (1998) and Paul (1982).

We first give the estimates by the method of moments. This method was first given by Kleinman (1973), who used weighted average and weighted variance of sample proportions to equate to their respective expected values to find method of moments estimates of the parameters π and ϕ.

Define $z_i = \frac{y_i}{n_i}$, $w_i = \frac{n_i}{\pi(1-\pi)\{1+(n_i-1)\phi\}}$ for $i = 1,2,\ldots,m$ and $\hat{\pi} = \sum_{i=1}^{m} w_i z_i / \sum_{i=1}^{m} w_i$. It can be seen that given ϕ, $\mathbf{E}(\hat{\pi}) = \pi$. Further, define $S = \sum_{i=1}^{m} w_i (z_i - \pi)^2$. Then it can be shown that

$$\mathbf{E}(S) = \sum_{i=1}^{m} \frac{w_i \pi (1-\pi)}{n_i} + \sum_{i=1}^{m} \frac{w_i \pi (1-\pi)(n_i - 1)\phi}{n_i}.$$

Then, the method of moments estimates of π and ϕ are obtained by solving the estimating equations

$$\sum_{i=1}^{m} w_i (z_i - \pi) = 0,$$

and

$$\sum_{i=1}^{m} w_i (z_i - \pi)^2 - \sum_{i=1}^{m} \frac{w_i \pi (1-\pi)}{n_i} - \sum_{i=1}^{m} \frac{w_i \pi (1-\pi)(n_i - 1)\phi}{n_i} = 0$$

simultaneously.

As is well known, the maximum likelihood estimates are obtained by maximizing the likelihood function with respect to the parameters of interest. Using model (7.22) the log-likelihood, apart from a constant, can be written as

$$l = \sum_{i=1}^{m} \left[\sum_{r=0}^{y_i-1} \log\{(1-\phi)\pi + r\phi\} + \sum_{r=0}^{n_i-y_i-1} \log\{(1-\phi)(1-\pi) + r\phi\} \right.$$
$$\left. - \sum_{r=0}^{n_i-1} \log\{(1-\phi) + r\phi\} \right], \tag{7.23}$$

and the maximum likelihood estimates of the parameters π and ϕ can be obtained by solving the estimating equations

$$\sum_{i=1}^{m} \left[\sum_{r=0}^{y_i-1} \frac{(1-\phi)}{(1-\phi)\pi + r\phi} - \sum_{r=0}^{n_i-y_i-1} \frac{(1-\phi)}{(1-\phi)(1-\pi) + r\phi} \right] = 0$$

and

$$\sum_{i=1}^{m} \left[\sum_{r=0}^{y_i-1} \frac{r-\pi}{(1-\phi)\pi + r\phi} + \sum_{r=0}^{n_i-y_i-1} \frac{r+\pi-1}{(1-\phi)(1-\pi) + r\phi} - \sum_{r=0}^{n_i-1} \frac{r-1}{(1-\phi) + r\phi} \right] = 0$$

simultaneously.

7.2.9   Inference

After the parameters are estimated, attention generally shifts to hypothesis testing and confidence interval construction. Recall that the beta-binomial and the extended beta-binomial models are extensions of the simpler binomial model. It is then natural to test whether the binomial model is good enough to model toxicological data in the form of proportions. To this end we test the null hypothesis $H_0 : \phi = 0$ against the alternative hypothesis $H_1 : \phi \neq 0$.

The pioneering work in this is by Tarone (1979) who developed $C(\alpha)$ (Neyman, 1959) tests for testing the goodness of the binomial model against the BB, CB and MB models. These tests are closely related to the binomial variance test. Paul (1982) summarized these results as follows.

Let $\hat{p} = y/n$, where $y = \sum_{i=1}^{m} y_i$ and $n = \sum_{i=1}^{m} n_i$ and $\hat{q} = 1 - \hat{p}$. Further, let $A = \sum_{i=1}^{m} n_i(n_i - 1)$, $B = \sum_{i=1}^{m} n_i(n_i - 1)^2$, $R = \sum_{i=1}^{m} y_i(n_i - y_i)$, $D = (\hat{q} - \hat{p})A$, $E = \hat{p}\hat{q}(B + \hat{p}\hat{q}(2A - 4B))$, $F = n/(\hat{p}\hat{q})$ and $s = \sum_{i=1}^{m}(y_i - n_i\hat{p})^2/(\hat{p}\hat{q})$. Then the optimal test statistic for testing the goodness of fit of the binomial distribution
(i) against the *BB* model is $Z = (s - n)/(2A)^{1/2}$, and under the null hypothesis that the binomial model has a good fit against *BB* model and under alternative hypothesis will have an asymptotic standard normal distribution;
(ii) against the CB model is $X_c^2 = Z^2$, and under the null hypothesis that the binomial model has a good fit against *CB* model, and under alternative hypothesis will have an asymptotic χ^2 distribution with one degree of freedom and;
(iii) against the MB model is $X_M^2 = (R - A\hat{p}\hat{q})^2(E - D^2/F)^{-1}$, and under the null hypothesis that the binomial model has a good fit against *MB* model alternative and under alternative hypothesis will have an asymptotic χ^2 distribution with one degree of freedom.

After an extensive simulation study to compare these three tests, Paul (1982) found that the BB model is, in general, more sensitive to the departure from the binomial model, and therefore, is a superior model for the analysis of the data in Table 7.2 given in Section 7.3. The justification for superiority of one model over the others is that a model which is more sensitive to the departure from the binomial will characterize the data more accurately than others which are less sensitive.

7.3   Some Examples

Data given in Paul (1982, Table 7.1 p 362) are obtained from a teratological experiment from Shell Toxicology Laboratory, Sittingbourne Research Centre, Sittingbourne, Kent, England. There were four groups consisting of a control (C) group, a low (L) dose group, a Medium (M) dose group, and a high (H) dose group.

By analyzing these data, Paul (1982) concluded that the beta-binomial model is the most sensitive to the departure from the binomial model for data sets of the type met in teratological experiments.

Below we give a subset of the data for interested readers to analyze.

Table 7.2 *Data from Toxicological experiment (Paul, 1982). (i) Number of live fetuses affected by treatment. (ii) Total number of live fetuses.*

Dose Group																			
Control (C)	(i) 1	1	4	0	0	0	0	0	1	0	2	0	5	2	1	2	0	0	1
	(ii) 12	7	6	6	7	8	10	7	8	6	11	7	8	9	2	7	9	7	11
Low (L)	(i) 0	1	1	0	2	0	1	0	1	0	0	3	0	0	1	5			
	(ii) 5	11	7	9	12	8	6	7	6	4	6	9	6	7	5	9			
Medium (M)	(i) 2	3	2	1	2	3	0	4	0	0	4	0	0	6	6	5			
	(ii) 4	4	9	8	9	7	8	9	6	4	6	7	3	13	6	8			
High (H)	(i) 1	0	1	0	1	0	1	1	2	0	4	1	1	4	2				
	(ii) 9	10	7	5	4	6	3	8	5	4	4	5	3	8	6				

7.4 Regression Models for Multilevel Clustered Data

Cluster-correlated data arise when data are obtained in clusters where data within clusters are correlated. Also data may be obtained where there is a natural structure or hierarchy within each cluster. Some examples are:

Example 1: In studies of health care services we may be interested in assessing quality of care for patients who are nested or grouped within different clinics. These data are said to be hierarchical data having two-levels: the patients within clinics are at level 1 and clinics are level 2 units.

Example 2: In family studies children are at level 1, mother-father at level 2 and families are at level 3.

Example 3: In educational studies children are nested within a classroom, classrooms are nested within schools and schools are nested within in a school district. Data obtained would be three level data.

Often it is of interest to analyze these multilevel data. Fitzmaurice, Laird, and Ware (2004) provide an excellent overview of multilevel data. Here we discuss models and inference procedures for two-level data.

7.5 Two-Level Linear Models

The two level linear model is given by

$$Y_{ij} = X_{ij}\beta + Z_{ij}b_j + e_{ij}, \qquad (7.24)$$

where Y_{ij} is the response on the ith unit at level 1 within the jth unit (cluster) at level 2, X_{ij} is a $1 \times p$ (row) vector of covariates, β is a $1 \times p$ (row) vector of fixed effects regression parameters, Z_{ij} is a design matrix for the random effects at level 2, b_j is the random effect at the jth unit (cluster) at level 2, and e_{ij} is the error. The random effects, b_j, vary across level 2 units but, for a given level 2 unit, are the same for all level 1 units. For example, Y_{ij} might be the outcome for the ith patient in the

jth clinic, where the clinics are a random sample of clinics from around the area. The random effects are assumed to be correlated across level 2 units, with mean zero and covariance $cov(b_j) = G$. The level 1 random components, e_{ij}, are assumed to be independent across level 1 units, with mean zero and variance $Var(e_{ij}) = \sigma^2$. In addition, the e_{ij}'s are assumed to be independent of the b_j's, with $Cov(e_{ij}, b_j) = 0$. That is, level 1 units are assumed to be conditionally independent given the level 2 random effects (and the covariates).

The regression parameters, β, are the fixed effects and describe the effects of covariates on the mean response

$$E(Y_{ij}) = X_{ij}\beta, \tag{7.25}$$

where the mean response is averaged over both level 1 and level 2 units. The two level model given by (7.24) also describes the effects of covariates on the conditional mean response given the random effect b_j as

$$E(Y_{ij}|b_j) = X_{ij}\beta + Z_{ij}b_j, \tag{7.26}$$

where the response is averaged over level 1 units only. The maximum likelihood estimates of the parameters β, σ^2, and G for longitudinal data when there is no missing responses are obtained in Chapter 9. Following these results we obtain maximum likelihood estimates of the parameters β, σ^2 and G of model (7.24) as

$$\hat{\beta} = \left(\sum_{i=1}^{N} X_i^T \Sigma_i^{-1} X_i\right)^{-1} \sum_{i=1}^{N} X_i^T \Sigma_i^{-1} y_i \tag{7.27}$$

$$\hat{\sigma}^2 = \frac{\sum_{i=1}^{N} e_i^T e_i}{\sum_{i=1}^{N} n_i}, \qquad \hat{G} = \sum_{i=1}^{N} \frac{b_i^T b_i}{N}, \tag{7.28}$$

where $e_i = y_i - X_i\beta - Z_i b_i$ (see Chapter 9). However, suppose the $b'_j s$ are assumed to be independent, $var(b_j) = \sigma_2^2$ and there is no covariate at the level 2 units. Since the data are clustered, observations within the same cluster are correlated, but are independent across clusters. Then, letting $Var(e_{ij}) = \sigma_1^2$, the degree to which the observations within the same cluster are correlated can be measured by the intra-cluster correlation

$$\rho = \frac{\sigma_2^2}{\sigma_1^2 + \sigma_2^2}. \tag{7.29}$$

7.6   An Example: Developmental Toxicity Study of Ethylene Glycol

We refer to the data in Price et al. (1985). Developmental toxicity studies of laboratory animals play a crucial role in the testing and regulation of chemicals and pharmaceutical compounds. Exposure to developmental toxicants typically causes a variety of adverse effects, such as fetal malformations and reduced fetal weight at term. In a typical developmental toxicity experiment, laboratory animals are assigned to increasing doses of a chemical or a test substance. Price et al. (1985) provided data from an developmental toxicity study of ethylene glycol (EG). Ethylene glycol is a high volume industrial chemical used in many applications. It is used as an antifreeze, as a solvent in the paint and plastic industries, and in the formulation of various types of inks. In a study of laboratory mice conducted through the National Toxicology Program (NTP), EG was administered at doses of 0, 750, 1500, or 3000 mg/kg/day to 94 pregnant mice or dams over the period of major organogenesis, beginning just after implantation. See Price et al. (1985) for additional details concerning the study design. Following sacrifice, fetal weight and evidence of malformations were recorded for each live fetus. Fitzmaurice, Laird, and Ware (2004) provided an analysis of the data. In the analysis of the data, they focused on the effects of dose on fetal weight. Summary statistics (ignoring clustering in the data) for fetal weight for the 94 litters (composed of total 1028 live fetuses) are presented in Table 17.1, p 451 of Fitzmaurice et al. (2004) . Fetal weight decreases monotonically with increasing dose, with the average weight ranging from 0.972 (gm) in the control group to 0.704 (gm) in the group administered the highest dose. The decrease in fetal weight is not linear in increasing dose, but is approximately linear in increasing $\sqrt{\text{dose}}$.

The data on fetal weight from this experiment are clustered, with observations on the fetuses (level 1 units) nested within dams/litters (level 2 units). The litter sizes range from 1 to 16. Let Y_{ij} denote the fetal weight of the ith live fetus from the jth litter, Fitzmaurice et al. (2004) considered the following model relating the fetal weight outcome to dose:

$$Y_{ij} = \beta_1 + \beta_2 d_j + b_j + e_{ij}, \qquad (7.30)$$

where $d_j = \sqrt{\text{dose}_j/750}$ is the square-root transformed dose administered to the jth dam. The random effect b_j is assumed to vary independently across litters, with $b_j \sim N(0, \sigma_2^2)$. The errors, e_{ij}, are assumed to vary independently across fetuses (within a litter), and $e_{ij} \sim N(0, \sigma_1^2)$. This model assumes that fetuses within a cluster are exchangeable and the positive correlation among the fetal weights is accounted for by their sharing a common random effect, b_j. The degree of clustering of the data can be expressed in terms of the intra-cluster (or intra-litter) correlation

$$\rho = \frac{\sigma_2^2}{\sigma_1^2 + \sigma_2^2}. \qquad (7.31)$$

In Table 17.2, Fitzmaurice et al. (2004, 0.452) give the results of fitting the model to the fetal weight data. The REML (restricted maximum likelihood) estimate of the regression parameter for (transformed) dose indicates that the mean fetal weight decreases with increasing dose. The estimated decrease in weight, comparing the highest dose group to the control group, is 0.27 (or 2×-0.134, with the 95% confidence interval: -0.316 to - 0.220). Note that both model based and empirical (or sandwich) standard errors are calculated and they were very similar, suggesting that the simple random effect structure for the clustering of fetal weights is adequate. The estimate of the intra-cluster correlation, $\hat{\rho} = 0.57$, indicates that there are moderate litter effects.

Fitzmaurice et al. (2004) provided further analysis of the data to assess the adequacy of the linear dose-response trend. They considered a model that included a quadratic effect of (transformed) dose. Both Wald and likelihood ratio tests of the quadratic effect of dose indicated that the linear trend is adequate for these data (Wald $W^2 = 1.38$, with 1 df, p-value > 0.20; likelihood ratio $G^2 = 1.37$, with 1 df, p-value > 0.20).

7.7 Two-Level Generalized Linear Model

As for the two level linear model, Fitzmaurice et al. (2004) provided an excellent review of two level generalized linear model.

Let Y_{ij} be the response on the ith level 1 unit in the jth level 2 cluster. Let X_{ij} be a $1 \times p$ (row) vector of covariances. The response can be continuous, binary, or a count. Three-part specification for the two level generalized linear models by McCullagh and Nelder (1989) is given as follows.

(1) We assume that the conditional distribution of each Y_{ij}, given a vector of random effects b_j (and the covariances), belongs to the exponential family of distributions and that $\text{Var}(Y_{ij}|b_j) = v\{\mathbf{E}(Y_{ij}|b_j)\}\phi$, where $v(\cdot)$ is a known variance function and is a function of the conditional mean, $\mathbf{E}(Y_{ij}|b_j)$; and ϕ is a scale or dispersion parameter. In addition, given the random effects b_j, it is assumed that the Y_{ij} are independent of one another.

(2) The conditional mean of Y_{ij} is assumed to depend upon the fixed and random effects via the following linear predictor,

$$\eta_{ij} = X_{ij}\beta + Z_{ij}b_j, \qquad (7.32)$$

with

$$g\{\mathbf{E}(Y_{ij}|b_j)\} = \eta_{ij} = X_{ij}\beta + Z_{ij}b_j, \qquad (7.33)$$

for some known link function, $g(\cdot)$.

(3) Finally, the random effects are assumed to have some probability distribution. In principle, any multivariate distribution can be assumed for the b_j; in practice, for computational convenience, the random effects are usually assumed to have a multivariate normal distribution, with zero mean and covariance matrix, G. It is also common, for convenience, to assume that the b_j are independent, $\text{Var}(b_j) = \sigma_j^2$ and $\text{Cov}(b_j, b_k) = 0$ for $j \neq k$.

These three components completely specify a broad class of two level generalized linear models for different types of responses.

The two-level generalized linear model is generally referred to as generalized linear mixed models (GLMMs, Jiang, 2007). Next, to clarify the main ideas, we consider two examples of two level generalized linear models.

Example 1: Two-level Generalized Linear Model for counts

Consider a study comparing cross-national rates of skin cancer and the factors (e.g., climate, economic and social factors, regional differences in diagnostic procedures) that influence variability in the rates of disease. Suppose we have counts of the number of cases of skin cancer in a set of well-defined regions, indexed by i, within counties, indexed by j. Let Y_{ij} be a count of the number of individuals who develop skin cancer within the ith region of the jth county during a given period of time (say, 5 years). The resulting counts have a two level structure with regional units at the lower level (level 1 units). Usually, the analysis of count data requires knowledge of the population at risk. That is, the rate at which the disease occurs is of more direct interest than the corresponding count.

Counts are often modeled as Poisson random variables using a log link function. This motivates the following illustration of a two level generalized linear model for Y_{ij} given by the three part specification:

(1) Conditional on a vector of random effects b_j, the Y_{ij} are assumed to be independent observations from a Poisson distribution, with $\text{Var}(Y_{ij}|b_j) = \mathbf{E}(Y_{ij}|b_j)$, (i.e., $\phi = 1$).

(2) The conditional mean of Y_{ij} depends upon fixed and random effects via the following log link function,

$$\log\{\mathbf{E}(Y_{ij}|b_j)\} = \log(T_{ij}) + X_{ij}\beta + Z_{ij}b_j, \tag{7.34}$$

where T_{ij} is the population at risk in the ith region of the jth county and the $\log(T_{ij})$ is an offset.

(3) The random effects are assumed to have a multivariate normal distribution, with zero mean and covariance matrix G.

This is an example of a two-level log linear model that assumes a linear relationship between the log rate of disease occurrence and the covariances.

Example 2: Two-level Generalized Linear Model for Binary Responses

Consider a study of men with newly diagnosed prostate cancer. The study is designed to evaluate the factors that determine physician recommendations for surgery (radical prostatectomy) versus radiation therapy. In particular, it is of interest to determine the relative importance of patient factors (e.g., patients age, level of prostate specific antigen) and physician factors (e.g., speciality training, years of experience) on physician recommendations for treatment. Many patients in the study seek the recommendation of the same physician. As a result, patients (level 1 units) are nested within the physicians (level 2 units). For each patient, we have a binary outcome denoting the physicians recommendation (surgery versus radiation therapy).

Let Y_{ij} be the binary response, taking values 0 and 1 (e.g., denoting surgery or radiation therapy) for the ith patient of the jth physician. An illustrative example of a two level logistic model for Y_{ij} is given by the following three part specification:

(1) Conditional on a single random effects b_j, the Y_{ij} are independent and have a Bernoulli distribution, with $\text{Var}(Y_{ij}|b_j) = \mathbf{E}(Y_{ij}|b_j)\{1 - \mathbf{E}(Y_{ij}|b_j)\}$, (i.e., $\phi = 1$).

(2) The conditional mean of Y_{ij} depends upon fixed and random effects via the following linear predictor:

$$\eta_{ij} = X_{ij}\beta + b_j, \tag{7.35}$$

and

$$\log\left\{\frac{Pr(Y_{ij}=1|b_j)}{Pr(Y_{ij}=0|b_j)}\right\} = \eta_{ij} = X_{ij}\beta + b_j. \tag{7.36}$$

That is, the conditional mean of Y_{ij} is related to the linear predictor by a logit link function.

(3) The single random effect b_j is assumed to have a univariate normal distribution, with zero mean and variance σ_1^2.

In this example, the model is a simple two-level logistic regression model with randomly varying intercepts.

7.8 Rank Regression

Finally, we consider rank regression for the linear model as in §6.2, $Y_i = \beta_0 \mathbf{1} + X_i\beta + \epsilon_i$, As in Jung and Ying (2003), we assume ϵ_{ik} are continuous random variables with the fundamental assumption that the median of any pairwise difference $\epsilon_{ik} - \epsilon_{jl}$ is 0.

Define the residual vector $e_i(\hat{\beta}) = Y_i - X_i\hat{\beta}$ for any vector $\hat{\beta}$ of dimension p. The components of $e_i(\hat{\beta})$ are $e_{ik}(\hat{\beta})$, $1 \le k \le n_i$.

To bypass modeling the correlations, Jung and Ying (2003) suggest to ignore the possible within-subject correlations and apply the usual rank estimation for β, i.e., minimizing the following loss function as given by (6.3), which leads to the following quasi score functions for β,

$$U_{JY}(\beta) = M^{-2}\sum_{i=1}^{N}\sum_{j=1}^{N}\sum_{k=1}^{n_i}\sum_{l=1}^{n_k}(x_{ik}-x_{jl})\mathrm{sgn}(e_{ik}-e_{jl}),$$

where $x_{ik}^{\top}$ is the k-th row of the design matrix X_i. This function is equivalent to that given by (6.3).

One feature in cluster data analysis is that the time order of the observations is often not important or not recorded, which is often the case in developmental studies. This often implies the exchangeable correlation may be appropriate. This also motivates us to avoid modeling the correlation structure by subsampling one observation from each cluster and then apply the classical rank method to the resulting N independent observations, $(y_i), 1 \le i \le N$, say. If the corresponding residuals are $(e_{(i)})$, the dispersion function is

$$M^{-2}\sum_{i=1}^{N}\sum_{j\ne i}^{N}|(e_{(i)})-(e_{(j)})|.$$

To make use of all the observations, we would repeat this resampling process many times. Conditional on sampling one observation from each cluster, the probability of y_{ij} being sampled is $1/n_i$. Therefore, the limiting dispersion function is

$$L_{DS}(\beta) = M^{-2}\sum_{i=1}^{N}\sum_{j\ne i}n_i^{-1}n_j^{-1}\sum_{k=1}^{n_i}\sum_{l=1}^{n_k}|e_{ik}-e_{jl}|.$$

This within-cluster resampling was first considered by Hoffman et al. (2001) for eliminating the bias when the cluster sizes are informative. This was further investigated by Williamson et al. (2003) in the context of the GEE setup. In the special case of comparing two groups, the L_{DS} becomes the testing statistic (S) proposed by Datta and Satten (2005).

This motivates us to consider the following weighted loss function for estimation of β,

$$L_W(\beta) = M^{-2} \sum_{i=1}^{N} \sum_{j \neq i} \sum_{k=1}^{n_i} \sum_{l=1}^{n_k} w_i w_j |e_{ik}(\beta) - e_{jl}(\beta)|, \tag{7.37}$$

where w_i is a pre-chosen nonnegative and bounded sequence to take account of the different cluster sizes. The corresponding quasi-score functions for β are

$$U_W(\beta) = M^{-2} \sum_{i=1}^{N} \sum_{j \neq i} w_i w_j \sum_{k=1}^{n_i} \sum_{l=1}^{n_k} (x_{ik} - x_{jl}) \operatorname{sgn}(e_{ik} - e_{jl}).$$

If we regard L_W as a generalization to L_{DS}, $w_i / (\sum_j w_j)$ can be regarded as the probability of sampling from cluster i. This probability is chosen to be inversely proportional to the cluster size for the within cluster sampling approach. It seems that sensible choices include (i) $w_i = 1$, (ii) $w_i = 1/n_i$, (iii) $w_i = 1/\sqrt{n_i}$; and (iv) $w_i = \{1 + (n_i - 1)\bar{\rho}\}^{-1}$, where $\bar{\rho}$ is the "average" within-cluster correlation. Wang and Zhao (2008) provided more details on simulation and asymptotic results.

7.8.1 National Cooperative Gallstone Study

In this study, 113 patients who had floating gallstones were randomly allocated to the high dose group (750 mg per day), 65 patients, or a placebo group, 48 patients. The serum cholesterol was measured for each patients at baseline and at months 6, 12, 20, and 24 of follow-up. Many cholesterol measurements were missing because patient follow-up was terminated (Wei and Lachin, 1984). The response Y_{ij} here were the increase in cholesterol levels at months 6, 12, 20, and 24 above the baseline value. One of the major interest in this study was to investigate the safety of the drug chenodiol for the treatment of cholesterol gallstones. Let G_i be the group indicator taking 0 for placebo and 1 for high dose. In this case, we treated time as a categorical variable which is expressed by three dummy variables T_{12}, T_{20}, and T_{24}. Let measurements at month 6 as the reference level. We consider the regression model,

$$Y_{ij} = \beta_0 + \beta_1 T_{12} + \beta_2 T_{20} + \beta_3 T_{24} + \beta_4 G_i + \epsilon_{ij}.$$

The method using $w_i = 1/n_i$ is refereed to as W_0. Using Jung and Ying's method, the estimates for $\beta_1, \beta_2, \beta_3$, and β_4 are 2.774, 12.647, 7.922, and 10.475, respectively. The hypothesis of no treatment or time effect can be constructed $H_0 : \beta_i = 0$, $i = 1, ..., 4$, respectively.

The average correlation $\bar{\rho}$ is estimated as 0.49, which is quite large. We, therefore, applied the weighted rank method with weights $\{1 + \bar{\rho}(n_i - 1)\}^{-1}$ (W_3) and obtained $\hat{\beta} = (3.143, 13.314, 7.751, 10.393)$.

Under H_0, $Q = \max\limits_{\{\beta_i=0\}} M U_W(\beta) \hat{V}_W^{-1} U_W(\beta)^\top$ has an asymptotic chi-square distribution with one degree of freedom (Jung and Ying, 2003). The corresponding Q values in weighted rank regression are obtained as 0.816, 13.864, 2.106, and 3.970 for β_i, $i = 1, ..., 4$, respectively. For comparison, the Q values from Jung and Ying's method are 0.653, 16.100, 2.425, and 3.833 correspondingly. The 95% confidence intervals for $\beta_1, \beta_2, \beta_3$, and β_4 are $(-3.176, 9.461)$, $(6.251, 20.376)$, $(-2.532, 18.034)$, and $(-0.057, 20.842)$. The one-sided hypotheses are available based on the variance estimates of $\hat{\beta}_W$ which may suggest a significant drug effect on increasing the serum cholesterol level.

7.8.2 Reproductive Study

This study was conducted to evaluate the effects of an experimental compound on general reproductive performance and postnatal measurements taken on pups. Altogether 30 dams were randomly assigned to three equal size treatment groups: control, a low dose and a high dose of the compound. There are only 7 litters which are available in the high dose group since one female did not conceive, one cannibalized her pups and one delivered a dead birth (Dempster et al., 1984).

Denote the low dose group as L (reference group) and the high dose group as H. The response variable Y is pup weights which are used to assess the treatment effects. For each dam, the numbers of pups, which will be included in our model as S, are different. An interesting complication is that the cluster size is informative since pups from smaller litters tend to have a higher weight than those from larger cluster. In addition, gender is also considered as a covariate denoted as G. Therefore, we have the following model:

$$Y_{ij} = \beta_0 + \beta_1 L + \beta_2 H + \beta_3 G + \beta_4 S + \epsilon_{ij}.$$

The estimates from the independence rank regression are $\hat{\beta}_{JY} = (-0.448, -0.936, -0.236, -0.123)^\top$, from which we also obtain an estimate of $\bar{\rho}$ as 0.456. The weighted rank regression using $w_i = \{1 + \bar{\rho}(n_i - 1)\}^{-1}$ produces the estimates of β as $\hat{\beta}_W = (-0.460, -0.886, -0.242, -0.126)^\top$. Except for gender, the estimates from these two methods give similar estimates to those from the random intercept model of Dempster et al. (1984) who obtained the estimates as $(-0.429, -0.859, -0.359, -0.129)^\top$. The standard deviation of individual parameter estimates $\hat{\beta}_W$ are $(0.196, 0.241, 0.083, 0.025)^T$, which are generally higher than those from the random effect model which are equal to $(0.150, 0.182, 0.047, 0.019)^T$ based on normality assumption. No violation of normality in the data set may be a reason for this. The hypothesis tests based on $\hat{\beta}_W$ conclude that all the variables in the model have statistically significant effects on the pup weights.

To investigate the impact of the informative cluster size on the regression models, we excluded litter size from the set of covariates. The estimates using weight $1/n_i$ are $(-0.445, -0.441, -0.148)^\top$. It is interesting to note that that similar estimates are obtained here except for the high dose group. Same conclusions can also be seen in the random effect model, where the estimates are $(-0.375, -0.355, -0.361)^T$. The hypothesis tests show that neither the low dose nor the high dose are significant in the random effect model. However, the normal value of -2.432 for testing the effect of the low dose in the weighted rank regression with weights $\{1 + \bar{\rho}(n_i - 1)\}^{-1}$ indicates that the low dose significantly affect pup wights.

The resistance of the weighted rank regression to informative cluster size will need more research to explore. However, our weighted method including Size as a covariate should produce more efficient estimators.

Chapter 8

Missing Data Analysis

8.1 Introduction

Missing data or missing values occur when no information is available on the response or some of the covariates or both the response and some covariates for some subjects who participate in a study of interest. There can be a variety of reasons for occurrence of missing values. Nonresponse occurs when the respondent does not respond to certain questions due to stress, fatigue or lack of knowledge. Some individuals in the study may not respond because some questions are sensitive. Missing data can create difficulty in analysis because nearly all standard statistical methods presume complete information for all the variables included in the analysis. Even a small number of missing observations can dramatically affect statistical analysis resulting in biased and inefficient parameter estimates and confidence intervals that are too wide or too short.

8.2 Missing Data Mechanism

How do missing values or observations occur? The way missingness occurs in a data set or the missing data mechanism can be divided into three parts, missing completely at random (MCAR), missing at random (MAR) and missing not at random (MNAR). See the pioneering work in this area by Rubin (1976) and Little and Rubin (1987). MCAR refers to the missingness that does not depend on observed as well as unobserved observations. For MCAR, probability of missingness is same for all the observations. For example, if answering a question depends on the result of a head after tossing a fair coin, then missingness of that answer is completely random. MAR refers to the missingness that depends only on observed values. In MAR, probability of missingness depends only on available observations not on unobserved observations. For example, missing information about age or income may depend on other available information, such as race and gender of the individual. When neither MCAR nor MAR hold, the data are said to be MNAR. In this, the probability of missingness depends on both observed and unobserved observations. For example, dropouts in medical studies are MNAR as some individuals in a study may not like the previous results and may be worried about future results of the study and drop out (see Rubin and Little, 1987 and Song, 2007).

DOI: 10.1201/9781315153636-8

To put things in proper perspective, using notation similar to Song (2007), let $y_i = (y_{i1}, y_{i2}, \ldots, y_{in_i})^T$ and

$$r_{ij} = \begin{cases} 1 & \text{if } y_{ij} \text{ is observed,} \\ 0 & \text{otherwise.} \end{cases} \tag{8.1}$$

As some of the observations in response may be missing, we write the response y_i as

$$y_i = \begin{cases} y_i^o & \text{if } y_i \text{ is observed,} \\ y_i^m & \text{if } y_i \text{ is missing.} \end{cases} \tag{8.2}$$

Further, let X_i, Z_i, and W_i be design matrices for fixed effects, random effects (if any), and missing data process, and let θ and ψ be vectors that parameterize the joint distribution of y_i and $r_i = (r_{i1}, r_{i2}, \ldots, r_{in_i})^T$, where $\theta = (\beta^T, \alpha^T)$ and ψ represents the measurement and missingness processes, respectively, β is the fixed effects parameter vector and α represents the variance components and/or association parameters. The full data (y_i, r_i) consist of the complete data and the missing data indicators.

Now, when data are incomplete due to a stochastic mechanism, the full data density is

$$f(y_i, r_i | X_i, Z_i, W_i, \theta, \psi),$$

which can be factorized as

$$f(y_i, r_i | X_i, Z_i, W_i, \theta, \psi) = f(y_i | X_i, Z_i, \theta) f(r_i | y_i, W_i, \psi). \tag{8.3}$$

Then, under an MCAR mechanism, the probability of an observation being missing is independent of the responses, and therefore

$$f(r_i | y_i, W_i, \psi) = f(r_i | y_i^o, y_i^m, W_i, \psi) = f(r_i | W_i, \psi),$$

so that

$$f(y_i, r_i | X_i, Z_i, W_i, \theta, \psi) \propto f(y_i | X_i, Z_i, \theta).$$

So, the data analysis is done based on only the complete cases.

Under MAR, the probability of an observation being missing is conditionally independent of the unobserved data given the values of the observed data, which implies that

$$f(r_i | y_i, W_i, \psi) = f(r_i | y_i^o, W_i, \psi).$$

So the model for data analysis is

$$f(y_i, r_i | X_i, Z_i, W_i, \theta, \psi) = f(y_i | X_i, Z_i, \theta) f(r_i | y_i^o, W_i, \psi), \tag{8.4}$$

and at the observed data level the model is

$$f(y_i^o, r_i | X_i, Z_i, W_i, \theta, \psi) = f(y_i^o | X_i, Z_i, \theta) f(r_i | y_i^o, W_i, \psi). \tag{8.5}$$

Under MNAR, the probability of an observation being missing depends on observed and unobserved data which implies that

$$f(r_i | y_i, W_i, \psi) = f(r_i | y_i^o, y_i^m, W_i, \psi).$$

Thus, at the observed data level, the model is

$$f(y_i^o, r_i | X_i, Z_i, W_i, \theta, \psi) = \int f(y_i^o | X_i, Z_i, \theta) f(r_i | y_i^o, y_i^m | W_i, \psi) dy_i^m. \qquad (8.6)$$

However, in practice, often, the above integral is intractable. So, some Monte Carlo method needs to be devised to replace the integral by summation.

8.3   Missing Data Patterns

The best description of missing data patterns is by Song (2007) which we follow here. To illustrate some common missing data patterns encountered in practice, consider the following hypothetical example adapted from Song (2007).

For example, a longitudinal study involves eight subjects, each having three visits. Half of them are randomized into the standard treatment and the other half into the new treatment. Blood pressure is the outcome variable of interest.

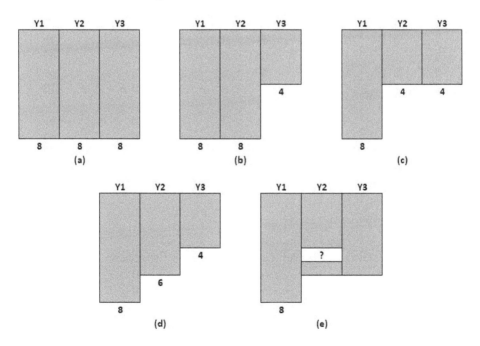

Figure 8.1 *Graphic representation of five missing data patterns*

A complete data pattern refers to the case with no missing values, as shown in Table 8.1 and in panel (a) of Figure 8.1. A univariate (response) missing pattern refers to the situation where missing values only occur at the last visit, as shown in Table 8.2 and panel (b) of Figure 8.1. This is a special case of dropout pattern.

Table 8.3 and panel (c) of figure 8.1 show a uniform missing pattern, in which missing values occur in a joint fashion. That is, the measurements at last two visits are either both observed or both missing simultaneously. This is a kind of dropout mechanism and the dropout time is uniform across all subjects.

Table 8.4 and panel (d) of Figure 8.1 display a monotonic missing pattern, where if one observation is missing, then all the observations after it will be unobserved. This is a general and important kind of dropout mechanism that allows subjects to have different dropout times. As a matter of fact, all the above cases (b)-(d) are monotonic missing patterns.

An arbitrary missing pattern refers to the case in which missing values may occur in any fashion, an arbitrary combination of intermittent missing values and dropouts. Table 8.5 and panel (e) of Figure 8.1 demonstrate a possible scenario for a mixture of intermittent missing (on Subject 5) at the second visit and some dropouts (on both Subject 1 and Subject 2).

Table 8.1 *Complete data pattern.*

Subject	Visit	Treatment	Heart Rate	Subject	Visit	Treatment	Heart Rate
1	1	A	65	5	1	B	72
1	2	A	64	5	2	B	74
1	3	A	61	5	3	B	76
2	1	A	60	6	1	B	67
2	2	A	80	6	2	B	69
2	3	A	81	6	3	B	73
3	1	A	62	7	1	B	62
3	2	A	68	7	2	B	61
3	3	A	69	7	3	B	65
4	1	A	81	8	1	B	75
4	2	A	75	8	2	B	74
4	3	A	78	8	3	B	72

Table 8.2 *Univariate missing data pattern.*

Subject	Visit	Treatment	Heart Rate	Subject	Visit	Treatment	Heart Rate
1	1	A	65	5	1	B	72
1	2	A	64	5	2	B	74
1	3	A	?	5	3	B	76
2	1	A	60	6	1	B	67
2	2	A	80	6	2	B	69
2	3	A	?	6	3	B	?
3	1	A	62	7	1	B	62
3	2	A	68	7	2	B	61
3	3	A	69	7	3	B	65
4	1	A	81	8	1	B	75
4	2	A	75	8	2	B	74
4	3	A	78	8	3	B	?

Table 8.3 *Uniform missing data pattern.*

Subject	Visit	Treatment	Heart Rate	Subject	Visit	Treatment	Heart Rate
1	1	A	65	5	1	B	72
1	2	A	?	5	2	B	74
1	3	A	?	5	3	B	76
2	1	A	60	6	1	B	67
2	2	A	?	6	2	B	?
2	3	A	?	6	3	B	?
3	1	A	62	7	1	B	62
3	2	A	68	7	2	B	61
3	3	A	69	7	3	B	65
4	1	A	81	8	1	B	75
4	2	A	75	8	2	B	?
4	3	A	78	8	3	B	?

Table 8.4 *Monotonic missing data pattern.*

Subject	Visit	Treatment	Heart Rate	Subject	Visit	Treatment	Heart Rate
1	1	A	65	5	1	B	72
1	2	A	64	5	2	B	74
1	3	A	?	5	3	B	76
2	1	A	60	6	1	B	67
2	2	A	?	6	2	B	69
2	3	A	?	6	3	B	?
3	1	A	62	7	1	B	62
3	2	A	68	7	2	B	61
3	3	A	69	7	3	B	65
4	1	A	81	8	1	B	75
4	2	A	75	8	2	B	?
4	3	A	78	8	3	B	?

8.4   Missing Data Methodologies

Missing data methodologies can be classified into imputational methodologies and likelihood but ad hoc procedures and the likelihood methodologies. The imputational methodologies are quick but inference based on this analysis can be misleading. The likelihood based procedures are much more intensive but formal data analytic procedures. The imputational procedures are given in Section 8.4.1 and the likelihood methods are discussed in Section 8.4.2. Section 8.6 deals with longitudinal data analysis with missing values.

Table 8.5 *Arbitrary missing data pattern.*

Subject	Visit	Treatment	Heart Rate	Subject	Visit	Treatment	Heart Rate
1	1	A	65	5	1	B	72
1	2	A	?	5	2	B	74
1	3	A	?	5	3	B	76
2	1	A	60	6	1	B	67
2	2	A	?	6	2	B	69
2	3	A	81	6	3	B	?
3	1	A	62	7	1	B	62
3	2	A	68	7	2	B	61
3	3	A	69	7	3	B	65
4	1	A	81	8	1	B	75
4	2	A	75	8	2	B	74
4	3	A	78	8	3	B	72

8.4.1   Missing Data Methodologies: The Methods of Imputation

8.4.1.1   Last Value Carried Forward Imputation

In this method, the last observed value of a subject is carried over to the next missing observation.

8.4.1.2   Imputation by Related Observation

Sometimes related observations are imputed to fill the missing values. It may happen in a study that the mother's age and educational status for a child is missing so the father's information is used to fill the mothers missing information.

8.4.1.3   Imputation by Unconditional Mean

In this type of imputation procedure, the missing value of a subject is replaced by the average of the available information of the same variable but from different subjects.

8.4.1.4   Imputation by Conditional Mean

This approach of imputation was discussed by Little and Rubin (1987). Following Molenberghs et al. (2004), conditional mean imputation can be explained by considering a single normal sample. The mean and the covariance matrix are calculated from the complete cases of the data in the first step, and then in the second step, information from the first step is used to calculate the conditional mean from a regression of missing values of a subject conditional on the actual observations. Conditional mean from the second step is used to replace the missing value.

8.4.1.5   Hot Deck Imputation

Hot deck imputation procedure uses similar responding units from the sample to replace the missing observations. This technique is one of the commonly used techniques. For example, if the information about the total number of persons in a household is missing then that information would be replaced by the total number of persons in a similar household in that area.

8.4.1.6   Cold Deck Imputation

In this imputation technique, missing observations are replaced with a constant value from the external sources, such as, a previous survey or study.

8.4.1.7   Imputation by Substitution

In this imputation technique, the missing observations or the nonresponses are substituted by the information from different sources or subjects which were not included initially in the survey. This method is usually used in the data collection stage of the survey. For example, if a previously selected subject was not found during the survey, then the information would be collected from another subject who was not selected initially in the survey.

8.4.1.8   Regression Imputation

In this method predicted values are obtained from regression of the missing observations on the observed values. For example, if the height and the weight are measured from 30 students of a class and the weight of a student was missing, then the weights of the 9 students would be regressed on their heights and the regression coefficients would be used for the prediction of the missing weight for that specific height.

8.4.2   Missing Data Methodologies: Likelihood Methods

Since the inception of the seminal paper by Rubin (1976) and the book by Little and Rubin (1987), a substantial amount of research (both theoretical and applied) have appeared in statistical literature (see for example, Carpenter et al., 2002, Chen and Ibrahim, 2002, Ibrahim et al., 2005 and Ibrahim and Molenberghs, 2009). Covering all topics in missing data literature is beyond the scope of this book. Here we give a general discussion of estimation based on likelihood methods and provide some examples of application.

In the usual situation where data are assumed from a distribution with a parameter θ, where θ can be vector valued, for example, normal (μ, σ^2), a likelihood L or a log-likelihood l is constructed to obtain maximum likelihood estimates (MLE) of the parameter. Then L or l is maximized to obtain MLE of θ. The situation becomes complicated when some observations are missing. In this situation the EM algorithm (Dempster et al., 1977) becomes handy. It is a general approach to iterative computation of maximum likelihood estimate when data are incomplete (or missing). The EM algorithm consists of an expectation step followed by a maximization step at each iteration which guarantees to converge to the MLE of the parameters. A cautionary note is that the current EM theory only guarantees convergence to a local maximum of the likelihood, which is not necessarily the MLE.

To put things in perspective, let $\mathbf{X}$ be a set of observed data and $\mathbf{Z}$ be a set of unobserved data from a probability distribution $f(y|\theta)$. Then the likelihood function can be written as $L(\theta; \mathbf{X}, \mathbf{Z}) = f(\mathbf{X}, \mathbf{Z}|\theta)$. The maximum likelihood estimate (MLE) of the unknown parameters θ is obtained by maximizing the marginal likelihood of the observed data

$$L(\theta; \mathbf{X}) = f(\mathbf{X}|\theta) = \sum_{\mathbf{Z}} f(\mathbf{X}, \mathbf{Z}|\theta).$$

Summation is replaced by integration for continuous data. However, in most situations summation is either difficult or intractable. The MLE from the marginal likelihood is then obtained by iteratively applying the following two steps:

Expectation step (E step): Calculate the expected value of the log likelihood function, with respect to the conditional distribution of $\mathbf{Z}$ given $\mathbf{X}$ under the current estimate of the parameters $\theta^{(t)}$:

$$Q(\theta|\theta^{(t)}) = \mathbf{E}_{\mathbf{Z}|\mathbf{X}, \theta^{(t)}} [\log L(\theta; \mathbf{X}, \mathbf{Z})],$$

Maximization step (M step): Find the parameter that maximizes this quantity:

$$\theta^{(t+1)} = \arg\max_{\theta} Q(\theta|\theta^{(t)}).$$

Example 1. (Modified data from Demster et al., 1977)

This is a motivating and interesting example adapted from Demster et al. (1977). The data used come from Rao (1965, pages 368-369) and refer to 207 animals that are distributed multinomially into four categories, so that the observed data consist of

$$y = (y_1, y_2, y_3, y_4) = (135, 18, 20, 34).$$

Rao (1965) used a genetic model for the population which specifies cell probabilities

$$\left(\frac{1}{2} + \frac{\pi}{4}, \frac{1-\pi}{4}, \frac{1-\pi}{4}, \frac{\pi}{4} \right)$$

for some π, $0 \le \pi \le 1$, so that the multinomial can be written as

$$f(y|\pi) = \frac{(y_1 + y_2 + y_3 + y_4)!}{y_1! y_2! y_3! y_4!} \left(\frac{1}{2} + \frac{\pi}{4} \right)^{y_1} \left(\frac{1-\pi}{4} \right)^{y_2} \left(\frac{1-\pi}{4} \right)^{y_3} \left(\frac{\pi}{4} \right)^{y_4}.$$

Rao (1965) reparameterized π by $\pi = (1-\theta)^2$. He then developed a Fisher-scoring procedure for maximizing $f(y|(1-\theta)^2)$ given the observed y.

Note that $P(Y_1 = y_1) = 1/2 + \pi/4$ which indicates that the first two categories have been put into one category having frequency 135. To use the EM algorithm we consider y as incomplete data from a five-category multinomial population having cell probabilities $1/2$, $\pi/4$, $(1-\pi)$, $(1-\pi)/4$, and $\pi/4$. So, the complete data consist of $x = (x_1, x_2, x_3, x_4, x_5)$, where $y_1 = x_1 + x_2$, $y_2 = x_3$, $y_3 = x_4$, $y_4 = x_5$, and the complete data likelihood is

$$f(x|\pi) = \frac{(x_1 + x_2 + x_3 + x_4 + x_5)!}{x_1! x_2! x_3! x_4! x_5!} \left(\frac{1}{2} \right)^{x_1} \left(\frac{\pi}{4} \right)^{x_2} \left(\frac{1-\pi}{4} \right)^{x_3} \left(\frac{1-\pi}{4} \right)^{x_4} \left(\frac{\pi}{4} \right)^{x_5}. \tag{8.7}$$

Note that the EM algorithm has two steps: the E-step and the M-step. Also x_1 and x_2 in (8.7) are unknown, and need to be estimated through the E-step. But we know that $x_1 + x_2 = 135$ and $P(X_1 = x_1) = \pi_1$ and $P(X_2 = x_2) = \pi_2$ with

$$\pi_1 = \frac{\frac{1}{2}}{\frac{1}{2} + \frac{1}{4}\pi} \quad \text{and} \quad \pi_2 = \frac{\frac{1}{4}\pi}{\frac{1}{2} + \frac{1}{4}\pi}.$$

Using the binomial (X_1, π_1), the expected values of X_1 and X_2 are $x_1 = 135\pi_1$ and $x_2 = 135\pi_2$.

The M step requires us to obtain the maximum likelihood estimate of π from (8.7). By taking the derivative of the log of (8.7) (log of the multinomial pdf) and equating it to zero, we obtain

$$\hat{\pi} = \frac{x_2 + 34}{x_2 + 18 + 20 + 34}.$$

The EM algorithm then proceeds as follows:

Let $\pi^{(p)}$ be the value π at the pth iteration. Then the value of x_2, denoted by $x_2^{(p)}$ (the E-step) is

$$x_2^{(p)} = 135 \frac{\frac{\pi^{(p)}}{4}}{\frac{1}{2} + \frac{\pi^{(p)}}{4}} \tag{8.8}$$

and the value of π at the $(p+1)$th iteration (M-step) is

$$\pi^{(p+1)} = \frac{x_2^{(p)} + 34}{x_2^{(p)} + 18 + 20 + 34}. \tag{8.9}$$

The MLE of π is obtained by cycling back and forth between (8.8) and (8.9).

It can be seen that starting with an initial value of $\pi^{(0)} = 0.5$, the algorithm converges in eight steps as can be seen in Table 8.6. By substituting $x_2^{(p)}$ from equation (8.8) into equation (8.9), and letting $\pi^{(*)} = \pi^{(p)} = \pi^{(p+1)}$ we can explicitly solve a quadratic equation for the maximum-likelihood estimate of π:

$$\pi^{(*)} = (25 + \sqrt{56929})/414 \doteq 0.6667101. \tag{8.10}$$

Example 2. (Casella and Berger, 2002)

Modeling incidence of a disease: Suppose that we observe a random sample $X_1, \ldots, X_n$ from Poisson (τ_i) and a random sample $Y_1, \ldots, Y_n$ from Poisson $(\beta\tau_i)$. This represents, for example, modeling incidence of a disease, Y_i, where the underlying rate is a function of an overall effect β and an additional factor τ_i. The parameter τ_i could be a measure of population density or health status of the population in area i. We do not see τ_i but get information on it through X_i.

Table 8.6 *The EM algorithm (Dempster et al., 1977)*

p	$\pi^{(p)}$	$\pi^{(p)} - \pi^{(*)}$	$(\pi^{(p+1)} - \pi^{(*)}) \div (\pi^{(p)} - \pi^{(*)})$
0	0.500000000	0.136710100	0.15031
1	0.616161600	0.020548500	0.13700
2	0.633895000	0.002815090	0.13517
3	0.636329600	0.000380520	0.1328
4	0.636658800	0.000051330	0.13492
5	0.636703200	0.00000693	0.13498
6	0.636709200	0.00000093	0.13420
7	0.636710000	0.00000013	0.13978
8	0.626821484	0 000000014	–

Suppose X_1 is missing and we want to estimate the parameters β and τ_i based on the remaining data. Now in case of no missing data, the joint distribution of (X_i, Y_i), $i = 1, \ldots, n$ is

$$f((x_1, y_1), \cdots, (x_n, y_n) | \beta, \tau_1, \tau_2, \cdots, \tau_n) = \prod_{i=1}^{n} \frac{e^{-\beta \tau_i} (\beta \tau_i)^{y_i}}{y_i!} \frac{e^{-\tau_i} (\tau_i)^{x_i}}{x_i!}. \qquad (8.11)$$

Then the maximum likelihood estimates are

$$\hat{\beta} = \frac{\sum_{i=1}^{n} y_i}{\sum_{i=1}^{n} x_i} \quad \text{and} \quad \hat{\tau}_j = \frac{x_j + y_j}{\hat{\beta} + 1}, \quad j = 1, 2, \ldots, n.$$

The joint distribution in (8.11) is the complete data likelihood, and $((x_1, y_1), \ldots, (x_n, y_n))$ is called the complete data. One way to handle estimation of the parameters when X_1 is missing is to delete Y_1 and estimate the parameters based on the $(n-1)$ pairs (X_i, Y_i), $i = 2, \ldots, n$. But this is ignoring the information in y_1. Using this information would improve our estimates.

Now, the likelihood of the sample with x_1 missing

$$\sum_{x_1=0}^{\infty} f((x_1, y_1), \ldots, (x_n, y_n) | \beta, \tau_1, \tau_2, \ldots, \tau_n) \qquad (8.12)$$

is the incomplete-data likelihood. This is the likelihood that we need to maximize. The incomplete data likelihood is obtained from (8.11) by summing over x_1. This gives

$$L(\beta, \tau_1, \tau_2, \cdots, \tau_n | (x_2, y_2), \ldots, (x_n, y_n)) = \prod_{i=1}^{n} \frac{e^{-\beta \tau_i} (\beta \tau_i)^{y_i}}{y_i!} \prod_{i=2}^{n} \frac{e^{-\tau_i} (\tau_i)^{x_i}}{x_i!} \qquad (8.13)$$

and $(y_1, (x_2, y_2), \ldots, (x_n, y_n))$ is the incomplete data (note that the summation over the likelihood (8.11) is equivalent to $\sum_{x_1=0}^{\infty} e^{-\tau_1} (\tau_1)^{x_1}$ which is 1). So we need to maximize the likelihood (8.13). Differentiation leads to the ML equations

$$\hat{\beta} = \frac{\sum_{i=1}^{n} y_i}{\sum_{i=1}^{n} \hat{\tau}_i}, \quad y_1 = \hat{\tau}_1 \hat{\beta}, \quad x_j + y_j = \hat{\tau}_j(\hat{\beta} + 1), \quad j = 2, \ldots, n. \qquad (8.14)$$

These equations can be solved iteratively.

We now use the EM algorithm to obtain the MLE's. Let $(\mathbf{x}, \mathbf{y}) = ((x_1, y_1),$ $\ldots, (x_n, y_n))$ denote the complete data and $(\mathbf{x}_{(-1)}, \mathbf{y}) = (y_1, (x_2, y_2), \ldots, (x_n, y_n))$ denote the incomplete data. The expected (expectation is over the missing data, that is x_1 here) complete data log likelihood is

$$E[\log L(\beta, \tau_1, \tau_2, \cdots, \tau_n | (\mathbf{x}, \mathbf{y}) | \tau^{(r)}, (\mathbf{x}_{(-1)}, \mathbf{y}))]$$

$$= \sum_{x_1=0}^{\infty} \log \left(\prod_{i=1}^{n} \frac{e^{-\beta\tau_i}(\beta\tau_i)^{y_i}}{y_i!} \frac{e^{-\tau_i}(\tau_i)^{x_i}}{x_i!} \right) \frac{e^{-\tau_1^{(r)}}(\tau_1^{(r)})^{x_1}}{x_1!}$$

$$= \sum_{i=1}^{n} [-\beta\tau_i + y_i(\log\beta + \log\tau_i) - \log y_i!] + \sum_{i=2}^{n} [-\tau_i + x_i \log\tau_i - \log x_i!]$$

$$+ \sum_{x_1=0}^{\infty} [-\tau_1 + x_1 \log\tau_1 - \log x_1!] \frac{e^{-\tau_1^{(r)}}(\tau_1^{(r)})^{x_1}}{x_1!}$$

$$= \left\{ \sum_{i=1}^{n} [-\beta\tau_i + y_i(\log\beta + \log\tau_i)] + \sum_{i=2}^{n} [-\tau_i + x_i \log\tau_i] \right.$$

$$\left. + \sum_{x_1=0}^{\infty} [-\tau_1 + x_1 \log\tau_1] \frac{e^{-\tau_1^{(r)}}(\tau_1^{(r)})^{x_1}}{x_1!} \right\}$$

$$- \left\{ \sum_{i=1}^{n} \log y_i! + \sum_{2=1}^{n} \log x_i! + \sum_{x_1=0}^{\infty} \log x_1! \frac{e^{-\tau_1^{(r)}}(\tau_1^{(r)})^{x_1}}{x_1!} \right\}, \qquad (8.15)$$

where in the last equality we have grouped together terms involving β and τ_i, and terms that do not involve these parameters. Since we are calculating this expected log likelihood for the purpose of maximizing it in β and τ_i, we can ignore the terms in the second set of parentheses, because

$$\sum_{i=1}^{n} \log y_i! + \sum_{2=1}^{n} \log x_i! + \sum_{x_1=0}^{\infty} \log x_1! \frac{e^{-\tau_1^{(r)}}(\tau_1^{(r)})^{x_1}}{x_1!}$$

is a constant. Thus, we have to maximize only the terms in the first set of parentheses. Now,

$$\sum_{x_1=0}^{\infty} [-\tau_1 + x_1 \log\tau_1] \frac{e^{-\tau_1^{(r)}}(\tau_1^{(r)})^{x_1}}{x_1!} = -\tau_1 + \log\tau_1 \sum_{x_1=0}^{\infty} x_1 \frac{e^{-\tau_1^{(r)}}(\tau_1^{(r)})^{x_1}}{x_1!} = -\tau_1 + \tau_1^{(r)} \log\tau_1.$$

Substituting this back into (8.15), apart from a constant, the expected complete data log-likelihood is

$$\sum_{i=1}^{n} [-\beta \tau_i + y_i(\log \beta + \log \tau_i)] + \sum_{i=1}^{n} (-\tau_i + x_i \log \tau_i).$$

This is the same as the original complete data log-likelihood, x_1 being replaced by $\tau_1^{(r)}$. Thus, in the rth step, the MLEs are only a minor variation of (8.14) and are given by

$$\hat{\beta}^{(r+1)} = \frac{\sum_{i=1}^{n} y_i}{\tau_1^{(r)} + \sum_{i=2}^{n} x_i} \quad \text{and} \quad \hat{\tau}_1^{(r+1)} = \frac{\tau_1^{(r)} + y_1}{\hat{\beta}^{(r+1)} + 1}, \quad \hat{\tau}_j^{(r+1)} = \frac{x_j + y_j}{\hat{\beta}^{(r+1)} + 1}, \quad j = 2, \dots, n.$$

(8.16)

This defines both the E-step (which results in the substitution of $\hat{\tau}_1^{(r)}$ for x_1) and the M-step (which results in the calculation in (8.16) for the MLEs at the rth iteration). Note that the final solution is obtained by substituting $\hat{\tau}_1^{(r)}$ on the first two equations in (8.16) by some initial value and then cycling back and forth.

Exercise 1. Refer to Example 2 above. Show that

(a) the maximum likelihood estimators from the complete data likelihood (8.11) are given by

$$\hat{\beta} = \frac{\sum_{i=1}^{n} y_i}{\sum_{i=1}^{n} x_i} \quad \text{and} \quad \hat{\tau}_j = \frac{x_j + y_j}{\hat{\beta} + 1}, \quad j = 1, 2, \dots, n.$$

and

(b) a direct solution of the original (incomplete-data) likelihood equations is possible. Show that the solution to (8.14) is given by

$$\hat{\beta} = \frac{\sum_{i=2}^{n} y_i}{\sum_{i=2}^{n} x_i}, \quad \hat{\tau}_1 = \frac{y_1}{\hat{\beta}}, \quad \hat{\tau}_j = \frac{x_j + y_j}{\hat{\beta} + 1}, \quad j = 2, 3, \dots, n. \tag{8.17}$$

Exercise 2. Use the model of Example 2 on the data in the following table adapted from Lange et al. (1994). These are leukemia counts and the associated populations for a number of areas in New York State.

Table 8.7 *Counts of leukemia cases*

population	3540	3560	3739	2784	2571	2729	3952	993	1908
Number of cases	3	4	1	1	3	1	2	0	2
Population	948	1172	1047	3138	5485	5554	2943	4969	4828
Number of cases	0	1	3	5	4	6	2	5	4

(a) Fit the Poisson model to these data both for the full data set and for an "incomplete" data set where we suppose that the first population count ($x_1 = 3540$) is missing.

(b) Suppose that instead of having an x value missing, we actually have lost a leukemia count (assume that $y_1 = 3$ is missing). Use the EM algorithm to find the MLEs in this case, and compare your answer to those of part (a).

8.5   Analysis of Zero-inflated Count Data With Missing Values

Discrete data in the form of counts often exhibit extra variation that cannot be explained by a simple model, such as the binomial or the Poisson. Also, these data sometimes show more zero counts than what can be predicted by a simple model. Therefore, a discrete model (Poisson or binomial) may fail to fit a set of discrete data either because of zero-inflation, or because of over-dispersion, or because there is zero-inflation as well as over-dispersion in the data. Deng and Paul (2005) developed score tests for zero-inflation and over-dispersion in generalized linear models. Mian and Paul (2016) developed estimation procedures for zero-inflated over-dispersed count data regression model with missing responses. Here we discuss these procedures in detail.

Let Y be a discrete count data random variable. The simplest model for such a random variable is the Poisson, which has the probability mass function

$$f(y;\mu) = \frac{e^{-\mu}\mu^{y}}{y!}. \tag{8.18}$$

However, data may show evidence of over-dispersion (variance is larger than the mean). A popular over-dispersed count data model is the two parameter negative binomial model. Different authors have used different parameterizations for the negative binomial distribution (see, for example, (Paul and Plackett, 1978; Barnwal and Paul, 1988; Paul and Banerjee, 1998; Piegorsch, 1990). Let Y be a negative binomial random variable with mean parameter μ and dispersion parameter c. Then, using the terminology of Paul and Plackett (1978), Y has the probability mass function

$$f(y;\mu,c) = \frac{\Gamma(y+c^{-1})}{y!\Gamma(c^{-1})} \left(\frac{c\mu}{1+c\mu}\right)^{y} \left(\frac{1}{1+c\mu}\right)^{c^{-1}}, \tag{8.19}$$

for $y = 0, 1, \ldots, \mu > 0$. Now, for a typical Y, $Var(Y) = \mu(1 + \mu c)$ and $c > -1/\mu$. This is the extended negative binomial distribution of Prentice (1986) which takes account of over-dispersion as well as under-dispersion. Obviously, when $c = 0$, variance of the $NB(\mu,c)$ distribution becomes that of the Poisson(μ) distribution. Moreover, it can be shown that the limiting distribution of the $NB(\mu,c)$ distribution, as $c \to 0$, is the Poisson(μ).

Using the mass function in equation (8.19), the zero-inflated negative binomial regression model (see Deng and Paul, 2005) can be written as

$$f(y_i|x_i;\mu,c,\omega) = \begin{cases} \omega + (1-\omega)\left(\frac{1}{1+c\mu}\right)^{c^{-1}} & \text{if } y = 0, \\ (1-\omega)\frac{\Gamma(y+c^{-1})}{y!\Gamma(c^{-1})}\left(\frac{c\mu}{1+c\mu}\right)^{y}\left(\frac{1}{1+c\mu}\right)^{c^{-1}} & \text{if } y > 0 \end{cases} \tag{8.20}$$

with $E(Y) = (1-\omega)\mu$, and $Var(Y) = (1-\omega)\mu[1+(c+\omega)\mu]$, where ω is the zero-inflation parameter. We denote this distribution as ZINB(μ,c,ω).

Regression analysis of count data may be further complicated by the existence of missing values either in the response variable and/or in the explanatory variables (covariates). Extensive work has been done on regression analysis of continuous response data with some missing responses under normality assumption. See, for example, Rubin (1976), Little and Rubin (1987), Anderson and Taylor (1976), Geweke (1986), Raftery et al. (1997), Chen et al. (2001), Kelly (2007), Zhang and Huang (2008).

Some work on missing values has also been done on logistic regression analysis of discrete data. See, for example, Ibrahim (1990), Lipsitz and Ibrahim (1996), Ibrahim and Lipsitz (1996), Ibrahim et al. (1999, 2001), Ibrahim et al. (2005), Sinha and Maiti (2008), Maiti and Pradhan (2009).

Exercise 3.

(a) Derive the negative binomial distribution as a *Gamma*(α,β) mixture of the Poisson(μ), reparameterize, and show that it can be written as the probability mass function given in equation (8.19) (see Paul and Plackett, 1978).

(b) Derive the mean and variance of the $NB(\mu,c)$ (hint: find the unconditional mean and unconditional variance of a mixture distribution).

(c) Verify that the mean and variance of a zero-inflated negative binomial distribution are those given in this chapter.

Suppose data for the ith of n subjects are (y_i, x_i), $i = 1,\ldots,n$, which are realizations from $ZINB(\mu,c,\omega)$, where y_i represents the response variable and x_i represents a $p \times 1$ vector of covariates with the regression parameter $\beta = (\beta_1,\beta_2,\ldots,\beta_p)$, such that $\mu_i = \exp(\sum_{j=1}^{p} X_{ij}\beta_j)$. Here β_1 is the intercept parameter in which case $X_{i1} = 1$ for all i. In Subsection 8.5.1 we show ML estimation of the parameters with no missing data. Subsection 8.5.2 deals with different scenarios of missingness.

8.5.1   Estimation of the Parameters with No Missing Data

For complete data the likelihood function is

$$
L(\beta,c,\omega|y_i) = \prod_{i=1}^{n} \left[\omega + (1-\omega)f(0;\mu_i,c,\omega)I_{\{y_i=0\}} + (1-\omega)f(y_i;\mu_i,c,\omega)I_{\{y_i>0\}} \right].
$$

(8.21)

Writing $\gamma = \omega/(1-\omega)$, the log likelihood, apart from a constant, can be written as

$$
l(\beta,c,\gamma|y_i) = \sum_{i=1}^{n} \left\{ -\log(1+\gamma) + \log[\gamma + f(0;\mu_i,c,\omega)]I_{\{y_i=0\}} + \log f(y_i;\mu_i,c,\omega)I_{\{y_i>0\}} \right\}
$$

$$
= \sum_{i=1}^{n} \left\{ -\log(1+\gamma) + \log\left[\gamma + \exp[-c^{-1}\log(1+\mu_i c)]\right]I_{\{y_i=0\}} \right.
$$

$$
\left. + \left[y_i\log\mu_i - (y_i+c^{-1})\log(1+\mu_i c) + \sum_{l=1}^{y_i}[1+(l-1)c] \right]I_{\{y_i>0\}} \right\}.
$$

(8.22)

The parameters β_j, c and γ can be estimated by directly maximizing the loglikelihood function (8.22) or by simultaneously solving the following estimating equations

$$\frac{\partial l}{\partial \beta_j} = \sum_{i=1}^{n} \left\{ \frac{-(1+\mu c)^{-1} \exp[-c^{-1} \log(1+\mu c)]}{\gamma + \exp[-c^{-1} \log(1+\mu c)]} I_{\{y_i=0\}} + \left[\frac{y_1}{\mu} - \frac{c(y_1 + c^{-1})}{1+\mu c} \right] I_{\{y_i>0\}} \right\}$$

$$\times \frac{\partial \mu_i}{\partial \beta_j} = 0, \tag{8.23}$$

$$\frac{\partial l}{\partial c} = \sum_{i=1}^{n} \left\{ \frac{[-\mu c^{-1}(1+\mu c)^{-1} + c^{-2} \log(1+\mu c)] \exp[-c^{-1} \log(1+\mu c)]}{\gamma + \exp[-c^{-1} \log(1+\mu c)]} I_{\{y_i=0\}} \right.$$

$$\left. + \left[\mu(y_i + c^{-1})(1+\mu c)^{-1} - c^{-2} \log(1+\mu c) + \sum_{l=1}^{y_i} (l-1) \right] I_{\{y_i>0\}} \right\} = 0, \tag{8.24}$$

and

$$\frac{\partial l}{\partial \gamma} = \sum_{i=1}^{n} \left\{ -(1+\gamma)^{-1} + \left[\gamma + \exp[-c^{-1} \log(1+\mu c)] \right]^{-1} I_{\{y_i=0\}} + 0 I_{\{y_i>0\}} \right\} = 0, \tag{8.25}$$

where

$$\frac{\partial \mu_i}{\partial \beta_j} = X_{ij} \exp\left(\sum_{j=1}^{p} X_{ij} \beta_j \right).$$

Exercise 4.

(a) Obtain maximum likelihood estimates of the parameters μ and c of model (8.19) for the modified DMFT index data given in Table 8A, 8B, 8C, 8D, 8E, 8F, 8G, 8H, 8I, and 8J.

(b) Obtain maximum likelihood estimates of the parameters μ, c and ω of model (8.20) for the modified DMFT index data given in Table 8A, 8B, 8C, 8D, 8E, 8F, 8G, 8H, 8I, and 8J.

8.5.2 Estimation of the Parameters with Missing Responses

8.5.2.1 Estimation under MCAR

In case of MCAR, missingness of the data does not depend on observed data and the subjects having the missing observations are deleted before the analysis. For estimation procedure, the likelihood function remains the same as given in equation (8.21) with reduced sample size having only complete observations. Note that this is the so-called complete-data analysis, which results in loss of information. For example, if every subject has at least one missing data component, which could be a response or a covariate, then, with the complete-data only analysis there will be no data left to analyze.

8.5.2.2   Estimation under MAR

As some of the observations in response may be missing, we write the response y_i as

$$y_i = \begin{cases} y_{o,i} & \text{if } y_i \text{ is observed,} \\ y_{m,i} & \text{if } y_i \text{ is missing.} \end{cases} \tag{8.26}$$

Using this in $f(y_i|x_i; \mu, c, \omega)$ given in equation (8.20), the log-likelihood is

$$l(\psi|Y_o, Y_m, X) = \sum_{i=1}^{n} \log(f(y_i|x_i, \psi))$$

$$= \sum_{i=1}^{n} \left\{ \log\left(\frac{\gamma + f(0; \mu_i, c, \omega)}{1 + \gamma}\right) I_{\{y_i=0\}} + \log f(y_i; \mu_i, c, \omega) I_{\{y_i>0\}} \right\}, \tag{8.27}$$

where Y_o is the vector of observed values, Y_m is the vector of missing values, $\psi = (\beta, c, \gamma)$ and $\mu_i = \exp(\sum_{j=1}^{p} X_{ij}\beta_j)$.

In MAR, conditional probability of missingness of the data depends on observed data. Parameters of the missingness mechanism are completely separate and distinct from the parameters of the model (8.20). In likelihood based estimation considering MAR, missingness mechanism can be ignored from the likelihood and missing data that are missing at random are often known as ignorable missing or ignorable non-response, but the subjects having these missing observations cannot be deleted before the analysis (see Little and Rubin (1987), Ibrahim et al. (2005) for detailed discussion on this).

In this scenario, our goal is to maximize the following loglikelihood with respect to the parameters ψ

$$l(\psi|Y_o, X) = \sum_{Y_m} l(\psi|Y_o, Y_m, X). \tag{8.28}$$

Note that $l(\psi|Y_o, X)$ is the log-likelihood when the missing data indicators, which are also part of the observed data, are not used.

In the more general case where missing data are not MAR, this likelihood would remain the same but a distribution defining the missing data mechanism needs to be included in the model. This general case is explained in the following section.

Direct maximization of $l(\psi; Y_o, X)$ is not, in general, straight forward. So, we use the EM algorithm.

As explained earlier, the EM algorithm uses an expectation-step (E-step) and a maximization-step (M-step). Following Little and Rubin (1987), the E-step provides the conditional expectation of the log-likelihood $l(\psi|y_{o,i}, y_{m,i}, x_i)$ given the observed data $(y_{o,i}, x_i)$ and current estimate of the parameters ψ.

Suppose A of the n responses are observed and $B = n - A$ responses are missing, and s is an arbitrary number of iterations during maximization of the log-likelihood. Then the E-step of the EM algorithm for the ith missing response for $(s + 1)$th iteration can be written as

$$Q_i(\psi|\psi^{(s)}) = \mathbf{E}[l(\psi^{(s)}|y_{o,i}, y_{m,i}, x_i)|y_{o,i}, x_i, \psi^{(s)}]$$

$$= \sum_{y_{m,i}} l(\psi^{(s)}|y_{o,i}, y_{m,i}, x_i)P(y_{m,i}|y_{o,i}, x_i, \psi^{(s)}). \tag{8.29}$$

For all the observations, the E-step of EM algorithm for $(s+1)$-th iteration is

$$Q(\psi|\psi^{(s)}) = \sum_{i=1}^{A} l(\psi^{(s)}|y_i) + \sum_{i=1}^{B} \sum_{y_{m,i}} l(\psi^{(s)}|y_{o,i}, y_{m,i}, x_i)P(y_{m,i}|y_{o,i}, x_i, \psi^{(s)}).$$

$$\tag{8.30}$$

Note that for the situation in which there is no missing response the EM algorithm requires only maximization of the first term on the right hand side.

Here $P(y_{m,i}|y_{o,i}, x_i, \psi^{(s)})$ is the conditional distribution of the missing response given the observed data and the current (s-th iteration) estimate of ψ. However, in many situations, $P(y_{m,i}|y_{o,i}, x_i, \psi^{(s)})$ may not always be available. Following Ibrahim et al. (2001) and Sahu and Roberts (1999), we can write $P(y_{m,i}|y_{o,i}, x_i, \psi^{(s)}) \propto P(y_i|x_i, \psi^{(s)})$ (the complete data distribution given in (8.20)). For the ith of the B missing responses, we take a sample $a_{i1}, a_{i2}, \ldots, a_{im_i}$ from $P(y_i|x_i, \psi^{(s)})$ using Gibbs sampler (see Casella and George (1992) for details). Then, following Ibrahim et al. (2001), $Q(\psi|\psi^{(s)})$ can be written as

$$Q(\psi|\psi^{(s)}) = \sum_{i=1}^{A} l(\psi^{(s)}|y_i) + \sum_{i=1}^{B} \frac{1}{m_i} \sum_{k=1}^{m_i} l(\psi^{(s)}|y_{o,i}, x_i, a_{ik}). \tag{8.31}$$

In the M-step, $Q(\psi|\psi^{(s)})$ is maximized. Here maximizing $Q(\psi|\psi^{(s)})$ is analogous to maximization of the complete data log likelihood where each incomplete response is replaced by m_i weighted observations. More details of the EM algorithm by method of weights can be found in Ibrahim (1990), Lipsitz and Ibrahim (1996), Ibrahim and Lipsitz (1996), Ibrahim et al. (1999, 2001), Ibrahim et al.(2005), Sinha and Maiti (2007), Maiti and Pradhan (2009).

Covariance matrix of the parameter estimators is calculated by inverting the observed information matrix at convergence (Efron and Hinkley, 1978), which is

$$H_{\psi\psi^T} = Q''(\psi|\psi^{(s)}) = \sum_{i=1}^{A} \frac{\partial^2}{\partial\psi\partial\psi^T} l(\psi^{(s)}|y_i) + \sum_{i=1}^{B} \frac{1}{m_i} \sum_{k=1}^{m_i} \frac{\partial^2 l(\psi^{(s)}|y_{o,i}, x_i, a_{ik})}{\partial\psi\partial\psi^T}. \tag{8.32}$$

Let $A(\mu,c) = \exp(-c^{-1}\log(1+\mu c))$. Expressions for the elements of H are

$$\frac{\partial^2 l}{\partial \beta_j^2} = \sum_{i=1}^{n} \frac{A(\mu,c)[\gamma(c+1)+cA(\mu,c)]}{(1+\mu c)^2[\gamma+A(\mu,c)]^2}\left(\frac{\partial \mu_i}{\partial \beta_j}\right)^2 I_{\{y_i=0\}}$$

$$- \sum_{i=1}^{n}\left[\frac{(1+\mu c)^{-1}\exp[-c^{-1}\log(1+\mu c)]}{\gamma+A(\mu,c)}I_{\{y_i=0\}} - \left(\frac{y_1}{\mu} - \frac{c(y_1+c^{-1})}{1+\mu c}\right)I_{\{y_i>0\}}\right]\frac{\partial^2 \mu_i}{\partial \beta_j^2}$$

$$+ \sum_{i=1}^{n}\left[\frac{c^2(y_i+c^{-1})}{(1+\mu c)^2} - \frac{y_i}{\mu^2}\right]\left(\frac{\partial \mu_i}{\partial \beta_j}\right)^2 I_{\{y_i>0\}},$$

$$\frac{\partial^2 l}{\partial \beta_j \partial \beta_j^{\mathrm{T}}} = \sum_{i=1}^{n} \frac{A(\mu,c)[\gamma(c+1)+cA(\mu,c)]}{(1+\mu c)^2[\gamma+A(\mu,c)]^2}\left(\frac{\partial \mu_i}{\partial \beta_j}\frac{\partial \mu_i}{\partial \beta_j^{\mathrm{T}}}\right)I_{\{y_i=0\}}$$

$$- \sum_{i=1}^{n}\left\{\frac{(1+\mu c)^{-1}A(\mu,c)}{\gamma+A(\mu,c)}I_{\{y_i=0\}} - \left(\frac{y_1}{\mu} - \frac{c(y_1+c^{-1})}{1+\mu c}\right)I_{\{y_i>0\}}\right\}\frac{\partial^2 \mu_i}{\partial \beta_j \partial \beta_j^{\mathrm{T}}},$$

$$+ \sum_{i=1}^{n}\left[\frac{c^2(y_i+c^{-1})}{(1+\mu c)^2} - \frac{y_i}{\mu^2}\right]\left(\frac{\partial \mu_i}{\partial \beta_j}\frac{\partial \mu_i}{\partial \beta_j^{\mathrm{T}}}\right)I_{\{y_i>0\}}$$

$$\frac{\partial^2 l}{\partial \beta_j \partial c} = \sum_{i=1}^{n} \frac{A(\mu,c)}{(1+\mu c)[\gamma+A(\mu,c)]^2}\left[\frac{\gamma \mu(1+c)+\mu cA(\mu,c)}{c(1+\mu c)} - \gamma c^{-2}\log(1+\mu c)\right]\frac{\partial \mu_i}{\partial \beta_j}I_{\{y_i=0\}}$$

$$- \sum_{i=1}^{n} \frac{c(1+\mu c)(y_i-c^{-2})+(1+2\mu c)(y_i+c^{-1})}{(1+\mu c)^2}\frac{\partial \mu_i}{\partial \beta_j}I_{\{y_i>0\}},$$

$$\frac{\partial^2 l}{\partial c^2} = \sum_{i=1}^{n} \frac{A(\mu,c)}{(\gamma+A(\mu,c))^2}\left\{\frac{\mu^2(1+\mu c)+2\mu c}{c^2(1+\mu c)^2} - 2c^{-3}\log(1+\mu c)\right\}I_{\{y_i=0\}}$$

$$+ \sum_{i=1}^{n} \frac{A(\mu,c)(1-A(\mu,c))}{(\gamma+A(\mu,c))^2}\left(\frac{\mu}{c(1+\mu c)} - c^{-2}\log(1+\mu c)\right)^2 I_{\{y_i=0\}}$$

$$+ \sum_{i=1}^{n}\left[-\frac{\mu^2(y_i+c^{-1})}{(1+\mu c)^2} - 2\frac{\mu}{c^2(1+\mu c)} + 2c^{-3}\log(1+\mu c)\right]I_{\{y_i>0\}},$$

and

$$\frac{\partial^2 l}{\partial c \partial \gamma} = \sum_{i=1}^{n}\left\{\frac{A(\mu,c)}{[\gamma+A(\mu,c)]^2}\left[\frac{\mu}{c(1+\mu c)} - c^{-2}\log(1+\mu c)\right]I_{\{y_i=0\}} + 0I_{\{y_i>0\}}\right\}$$

$$\frac{\partial^2 l}{\partial \beta_j \partial \gamma} = \sum_{i=1}^{n}\left[\frac{\exp(-c^{-1}\log(1+\mu c))}{(1+\mu c)[\gamma+A(\mu,c)]^2}\frac{\partial \mu_i}{\partial \beta_j}I_{\{y_i=0\}} + 0I_{\{y_i>0\}}\right],$$

$$\frac{\partial^2 l}{\partial \gamma^2} = \sum_{i=1}^{n}\left[(1+\gamma)^{-2} - [\gamma+A(\mu,c)]^{-2}I_{\{y_i=0\}} + 0I_{\{y_i>0\}}\right].$$

8.5.2.3 Estimation under MNAR

Under MNAR, probability of missing observations in the response variable depends on the covariates and the values of the response that would have been observed. Then, it is necessary to incorporate this missing data mechanism into the likelihood. The missing observations that follow this missing data mechanism are known as nonignorable missing. To incorporate the missing data mechanism into the likelihood we define a a binary random variable r_i $(i = 1, 2, \ldots, n)$ as

$$r_i = \begin{cases} 0 & \text{if } y_i \text{ is observed,} \\ 1 & \text{if } y_i \text{ is missing.} \end{cases} \tag{8.33}$$

Obviously, as a binary random variable, r_i follows

$$p(r_i | y_i, x_{ij}) = [p(r_i = 1)]^{r_i} [1 - p(r_i = 1)]^{(1-r_i)}. \tag{8.34}$$

See Ibrahim et al. (2001). To model the probability of missing in terms of the values of the responses that would have been observed and the covariates, a logit link,

$$\log\left[\frac{p(r_i = 1)}{1 - p(r_i = 1)}\right] = v_0 + v_1 * y_i + v_2 * x_{i1} + v_3 x_{i2} + \cdots + v_{p+1} x_{ip}, \tag{8.35}$$

can be used, where y_i is the responses and the responses that would have been observed, and x_{ij} $(j = 1, 2, \ldots, p)$ are the covariates. Denote the $(p+2)$ parameter vector as $v = (v_0, v_1, v_2, \ldots, v_{p+1})$. Note that $p(r_i = 1)$ can now be written as a logistic model

$$p(r_i = 1) = \frac{\exp(v_0 + v_1 * y_i + v_2 * x_{i1} + v_3 x_{i2} + \cdots + v_q x_{ip})}{1 + \exp(v_0 + v_1 * y_i + v_2 * x_{i1} + v_3 x_{i2} + \cdots + v_{p+1} x_{ip})}. \tag{8.36}$$

Then, the log-likelihood function of the parameter v can be written as

$$l(v | r_i, y_i, x_{ij}) = \sum_{i=1}^{n} \left\{ r_i * \log\left[\frac{p(r_i = 1)}{1 - p(r_i = 1)}\right] + \log(1 - p(r_i = 1)) \right\}. \tag{8.37}$$

Note that choice of variables for the model of r_i is important. Often many variables in this model are not necessarily significant, and more importantly parameters in the model for r_i are not of primary interest. Detailed discussion on this can be found in Ibrahim, Lipsitz, and Chen (1999) and Ibrahim, Chen, and Lipsitz (2001).

Following Ibrahim, Lipsitz, and Chen (1999), after incorporating the model for missingness mechanism in $l(v | r_i, y_i, x_{ij})$, the log-likelihood for all the parameters involved is

$$l(\psi | Y, X_{o,j}, X_{m,j}) = \sum_{i=1}^{n} \left\{ -\log(1 + \gamma) \right.$$

$$+ \log[\gamma + f(0; \mu_i, c, \omega)] I_{\{y_i = 0\}} + \log f(y_i; \mu_i, c, \omega) I_{\{y_i > 0\}} \right\}$$

$$+ \sum_{i=1}^{n} \left\{ r_i * \log\left[\frac{p(r_i = 1)}{1 - p(r_i = 1)}\right] + \log(1 - p(r_i = 1)) \right\}. \tag{8.38}$$

Note that two parts of this log-likelihood are separate and their parameters are distinct. This characteristic of the log-likelihood facilitates separate maximization. The rest of the estimation procedure under MNAR remains exactly the same as the estimation procedure under MAR. Note that as it is, the log-likelohood in (8.38) is not computable. However, it is computable when this is plugged in into the EM algorithm. Further, note that some of the covariates x_{ij} also may be missing which has not been discussed here as it needs separate theoretical development. This has been left for future research.

The theoretical development here is flexible. If through tests (Deng and Paul, 2005) or other data visualization procedures no evidence of zero-inflation or over-dispersion is evident in the data, then we can start with a simple model, such as the Poisson model or the negative binomial model or the zero-inflated Poisson model. Note that the development of missing data involved here is that of response data y_i. However, some of the covariates x_{ij} can also be missing which has not been dealt with here.

Example 3.

Mian and Paul (2016) analyzed a set of data from a prospective study of dental status of school children from Bohning et al. (1999). Here we report the results of the analysis. For more detailed information see, Mian and Paul (2016). The children were all 7 years of age at the beginning of the study. Dental status were measured by the decayed, missing and filled teeth (DMFT) index. Only the eight deciduous molars were considered so the smallest possible value of the DMFT index is 0 and the largest is 8. The prospective study was for a period of two years. The DMFT index was obtained at the beginning of the study and also at the end of the study.

The data also involved three categorical covariates: gender having two categories (0 - female, 1 - male), ethnic group having three categories (1 - dark, 2 - white, 3 - black) and school having six categories (1 - oral health education, 2 - all four methods together, 3 - control school (no prevention measure), 4 - enrichment of the school diet with ricebran, 5 - mouthrinse with 0.2% NaF-solution, 6 - oral hygiene).

For the purpose of illustration of estimation in a zero-inflated over-dispersed count data model with missing responses, Mian and Paul (2016) use the DMFT index data obtained at the beginning of the study (as in Deng and Paul, 2005). The DMFT index data at the beginning of the study are: (index, frequency): (0,172), (1,73), (2,96), (3,80), (4,95), (5,83), (6,85), (7,65), (8,48). They first fit a zero-inflated negative binomial model to the complete data and data with missing observations without covariates. Data with missing observations were obtained by randomly deleting a certain percentage (5%, 10%, 25%) of the observed responses.

The estimates of the mean parameter μ, the over dispersion parameter c and the zero inflation parameter ω based on the zero-inflated negative binomial model, under different percentages of missingness, and their corresponding standard errors are given in Table V of Mian and Paul (2016). Note that the estimates of the parameters μ, c and ω and the corresponding standard errors remain stable irrespective of the amount of missingness. Only for MAR and MNAR and for 25% missing values, their values are slightly different (slightly larger in case of μ and c). For higher percentage of missing these properties might deteriorate further.

For more insight $\widehat{\mathbf{E}(Y)} = (1 - \hat{\omega})\hat{\mu}$ and $\widehat{\mathrm{Var}(Y)} = (1 - \hat{\omega})\hat{\mu}[1 + (\hat{c} + \hat{\omega})\hat{\mu}]$ were calculated and are given in Table V of Mian and Paul (2016). These estimates also do not vary very much irrespective of the amount of missingness, except under MNAR and 25% missing, when $\widehat{\mathrm{Var}(Y)}$ is slightly higher compared to others.

Mian and Paul (2016) then fit a zero-inflated negative binomial model to the complete data and data with missing observations and covariates. Response data with missingness were obtained exactly the same way as in the situation without covariates. The model fitted was $\mu = \exp\{\beta + \beta_{G(M)}I(Gender = 1) + \beta_{E(D)}I(Ethnic = 1) + \beta_{E(W)}I(Ethnic = 2) + \beta_{S(1)}I(School = 1) + \beta_{S(2)}I(School = 2) + \beta_{S(3)}I(School = 3) + \beta_{S(4)}I(School = 4) + \beta_{S(5)}I(School = 5)\}$, where β represents the intercept parameter and β_G represents the regression parameter for gender, $\beta_{E(1)}$ and $\beta_{E(2)}$ represent the regression parameters for the ethnic groups 1 and 2, and $\beta_{S(1)}, \beta_{S(2)}, \beta_{S(3)}, \beta_{S(4)}$, and $\beta_{S(5)}$ represent the regression parameters for school 1, school 2, school 3, school 4, and school 5, respectively. Estimates of the parameters can be found in Table VI of Mian and Paul (2016).

In this case the estimates differed (this is expected as it depends on which observations have remained in the final data set). In general, the standard errors of the estimates are larger (in some cases these are much larger, for example, in case of $SE(\hat{\beta}_{S(5)})$) than those under complete data. For MCAR, MAR, and MNAR, and 25% missing responses, the standard error is close to twice for missing data in comparison to those for complete data. The estimates of $\widehat{\mathbf{E}(Y)}$ do not vary much irrespective of the percentage missing and the missing data mechanism. For complete data and smaller percentage missing, the property of $\widehat{\mathrm{Var}(Y)}$ is similar to that of $\widehat{\mathbf{E}(Y)}$. However, for MAR and 25% missing values $\widehat{\mathrm{Var}(Y)}$ is much larger (12.415) than in the other cases (varies between 8.26 and 9.5).

Here we create a subset of the DMFT data analysed by Mian and Paul (2016). These data are given in the Appendix under title: DMFT data for the analysis in chapter 8 (Table 8A, Table 8B, Table 8C, Table 8D, Table 8E, Table 8F, Table 8G, Table 8H, Table 8I, Table 8J, Table 8K). Further, we analyzed the modified DMFT data and the Results are given in Table 8.8 and Table 8.9. The analyses and conclusion of the results in Tables 8.8 and 8.9 are very similar to those given in Table V and Table VI in Mian and Paul (2016).

8.6   Analysis of Longitudinal Data With Missing Values

8.6.1   Normally Distributed Data

The concept of longitudinal data has been introduced earlier in this book. Here we follow Ibrahim and Molenberghs (2009) who provided an excellent review of missing data methods in longitudinal studies. We start with the normal random-effects model (Laird and Ware, 1982) with missing data (MAR or MNAR)

$$y_i = X_i\beta + Z_ib_i + e_i, \qquad\qquad i = 1,\ldots,N, \qquad\qquad (8.39)$$

where y_i is $n_i \times 1$, X_i is an $n_i \times p$ known matrix of fixed-effects covariates, β is a $p \times 1$ vector of unknown regression parameters, commonly referred to as fixed effects, Z_i

Table 8.8 *Estimates and standard errors of the parameters for DMFT index data.*

Percentage missingness		$\hat{\mu}$	$SE(\hat{\mu})$	$\hat{c}$	$SE(\hat{c})$	$\hat{\omega}$	$SE(\hat{\omega})$	$\widehat{E(y)}$	$\widehat{Var(y)}$
Complete data	0%	4.1444	0.0916	0.0426	0.0195	0.1892	0.0150	3.3603	6.5887
MCAR	5%	4.1598	0.0942	0.0415	0.0199	0.1953	0.0155	3.3473	6.6445
	10%	4.1440	0.0976	0.0472	0.0210	0.1922	0.0159	3.3475	6.6685
	25%	4.1514	0.1066	0.0465	0.0230	0.1878	0.0173	3.3719	6.6514
MAR	5%	4.1739	0.0891	0.0291	0.0182	0.1862	0.0148	3.3967	6.4497
	10%	4.1730	0.0872	0.0224	0.0175	0.1743	0.0144	3.4457	6.2731
	25%	4.2395	0.0846	0.0127	0.0160	0.1500	0.0135	3.6035	6.0887
MNAR	5%	4.1562	0.0939	0.0397	0.0197	0.1945	0.0154	3.3476	6.6071
	10%	4.1488	0.0976	0.0473	0.0210	0.1934	0.0160	3.3464	6.6878
	25%	4.1532	0.1069	0.0479	0.0232	0.1877	0.0173	3.3735	6.6742

Table 8.9 *Estimates and standard errors of the parameters for DMFT index data with covariates.*

Percentage missingness		$\hat{\beta}$	$SE(\hat{\beta})$	$\hat{\beta}_G$	$SE(\hat{\beta}_G)$	$\hat{\beta}_{E(1)}$	$SE(\hat{\beta}_{E(1)})$	$\hat{\beta}_{E(2)}$	$SE(\hat{\beta}_{E(2)})$
Complete data	0%	1.3000	0.0891	0.2551	0.0462	−0.0496	0.0731	0.0598	0.0688
MCAR	5%	1.5621	0.1006	0.0738	0.0495	−0.1391	0.0818	−0.0140	0.0760
	10%	1.1953	0.1068	0.3302	0.0554	−0.1125	0.0869	0.1454	0.0815
	25%	1.1401	0.1324	0.4094	0.0786	−0.2894	0.1106	0.2243	0.1031
MAR	5%	1.4724	0.1073	0.1756	0.0534	−0.2948	0.0932	0.0998	0.0813
	10%	1.3841	0.1258	0.2629	0.0656	−0.1999	0.1161	−0.1479	0.1040
	25%	1.3072	0.1164	0.5014	0.0730	−0.2073	0.0967	−0.1625	0.0901
MNAR	5%	1.5621	0.1006	0.0738	0.0495	−0.1391	0.0818	−0.0140	0.0760
	10%	1.1953	0.1068	0.3302	0.0553	−0.1125	0.0870	0.1454	0.0815
	25%	1.1401	0.1324	0.4094	0.0786	−0.2894	0.1106	0.2243	0.1031

Percentage missingness		$\hat{\beta}_{S(1)}$	$SE(\hat{\beta}_{S(1)})$	$\hat{\beta}_{S(2)}$	$SE(\hat{\beta}_{S(2)})$	$\hat{\beta}_{S(3)}$	$SE(\hat{\beta}_{S(3)})$	$\hat{\beta}_{S(4)}$	$SE(\hat{\beta}_{S(4)})$
Complete data	0%	0.1252	0.0739	−0.0955	0.0861	−0.1308	0.0788	−0.1093	0.0824
MCAR	5%	0.0372	0.0822	−0.2767	0.0992	−0.3295	0.0932	−0.2309	0.0943
	10%	0.1276	0.0882	0.0712	0.0975	−0.1676	0.0945	−0.0338	0.0948
	25%	−0.0873	0.1184	0.0786	0.1244	−0.3042	0.1325	0.1355	0.1212
MAR	5%	0.0716	0.0901	−0.1583	0.1045	−0.2423	0.0995	−0.0857	0.0976
	10%	−0.1402	0.1150	0.1079	0.1150	−0.3748	0.1401	0.1177	0.1056
	25%	−0.0585	0.1040	−0.1015	0.1145	−0.3191	0.1152	−0.0719	0.1108
MNAR	5%	0.0372	0.0822	−0.2767	0.0992	−0.3295	0.0932	−0.2309	0.0943
	10%	0.1276	0.0882	0.0712	0.0975	−0.1676	0.0945	−0.0338	0.0948
	25%	−0.0873	0.1184	0.0786	0.1244	−0.3042	0.1325	0.1355	0.1212

Percentage missingness		$\hat{\beta}_{S(5)}$	$SE(\hat{\beta}_{S(5)})$	$\hat{c}$	$SE(\hat{c})$	$\hat{\omega}$	$SE(\hat{\omega})$	$\widehat{E(y)}$	$\widehat{Var(y)}$
Complete data	0%	−0.1425	0.0823	0.0398	0.0203	0.1413	0.0132	3.4502	5.9609
MCAR	5%	−0.1504	0.0900	0.0873	0.0300	0.1544	0.0155	3.3980	6.7001
	10%	−0.3075	0.1015	0.0986	0.0322	0.1731	0.0176	3.2278	6.6513
	25%	−0.0492	0.1193	0.2154	0.0759	0.1538	0.0242	3.2800	7.9741
MAR	5%	−0.3911	0.1109	0.1594	0.0470	0.1567	0.0183	3.3008	8.0543
	10%	0.0105	0.1090	0.2525	0.1309	0.1598	0.0295	3.2994	8.0491
	25%	−0.0524	0.1047	0.1501	0.0479	0.2043	0.0219	3.0300	7.1204
MNAR	5%	−0.1504	0.0900	0.0873	0.0300	0.1544	0.0155	3.3980	6.7001
	10%	−0.3075	0.1015	0.0985	0.0322	0.1731	0.0176	3.2278	6.6513
	25%	−0.0492	0.1193	0.2154	0.0759	0.1538	0.0242	3.2800	7.9741

is a known $n_i \times q$ matrix of covariates for the $q \times 1$ vector of random effects b_i, and e_i is an $n_i \times 1$ vector of errors. The columns of Z_i are usually a subset of X_i, allowing for fixed effects as well as random intercepts and/or slopes. It is typically assumed that the e_i are independent, the b_i are i.i.d., the b_i are independent of the e_i, and

$$e_i \sim N_{n_i}(0, \sigma^2 I_{n_i}), \qquad b_i \sim N_q(0, D), \qquad (8.40)$$

where I_{n_i} is the $n_i \times n_i$ identity matrix, and $N_q(\mu, \Sigma)$ denotes the q-dimensional multivariate normal distribution with mean μ and covariance matrix Σ. The positive definite matrix D is the covariance matrix of the random effects and is typically assumed to be unstructured and unknown. However, in practice, if one is convinced that a simpler correlation structure, such as the exchangeable correlation matrix (see Zhang and Paul, 2013), is sufficient then the estimation procedure might be simpler. Under these assumptions, the conditional model, where conditioning refers to the random effects, takes the form

$$(y_i | \beta, \sigma^2, b_i) \sim N_{n_i}(X_i\beta + Z_i b_i, \sigma^2 I_{n_i}). \qquad (8.41)$$

The model in (8.41) assumes a distinct set of regression coefficients for each individual once the random effects are known. Integrating over the random effects, the marginal distribution of y_i is obtained as

$$(y_i | \beta, \sigma^2, D) \sim N_{n_i}(X_i\beta, Z_i D Z_i^{\mathrm{T}} + \sigma^2 I_{n_i}). \qquad (8.42)$$

For the MLE of the parameters of the model (8.42) where there is no missing data, see Laird and Ware (1982) and Jennrich and Schluchter (1986). Here, as in Section 8.5 and Ibrahim and Molenberghs (2009), we first show how to use the EM algorithm to estimate parameters of this model for complete data. Although the EM algorithm is not necessary here, it simplifies the estimation procedure.

8.6.2   Complete-data Estimation via the EM

In model (8.39), denote the observed data, the covariance parameters, and the complete data by $Y = (y_1, \ldots, y_N)$, $\theta = (\sigma^2, D)$ and $V = (y_1, b_1, \ldots, y_N, b_N)$, respectively. Then, the likelihood of β and θ given Y is

$$L(\beta, \theta | Y) = \prod_{i=1}^{N} f(y_i | \beta, \sigma^2, D), \qquad (8.43)$$

and the likelihood of β and θ given V is

$$L(\beta, \theta | V) = \prod_{i=1}^{N} f(y_i, b_i | \beta, \sigma^2, D) = \prod_{i=1}^{N} f(y_i | \beta, \sigma^2, b_i) f(b_i | D). \qquad (8.44)$$

As in Section 8.4, the EM algorithm has the E-step and the M-step. Because we are dealing with a random effects model, the M-step itself needs two steps. Now, the log-likelihood, apart from a constant, based on the observed data can be expressed as

$$\ell(\beta,\sigma^2,D) = -\frac{1}{2}\sum_{i=1}^{N}\log|\Sigma_i| - \frac{1}{2}\sum_{i=1}^{N}(y_i - X_i\beta)^{\mathrm{T}}\Sigma_i^{-1}(y_i - X_i\beta), \qquad (8.45)$$

where $\Sigma_i = Z_iDZ_i^{\mathrm{T}} + \sigma^2 I_{n_i}$. Then, given $\theta = (\sigma^2, D)$ and $V = (y_1, b_1, \ldots, y_N, b_N)$, the ML estimates of β obtained by equating $\frac{d\ell}{d\beta}$ to zero and solving for β are

$$\hat{\beta} = \left(\sum_{i=1}^{N} X_i^{\mathrm{T}}\Sigma_i^{-1}X_i\right)^{-1}\sum_{i=1}^{N}X_i^{\mathrm{T}}\Sigma_i^{-1}y_i. \qquad (8.46)$$

This is the M-step for estimating β given θ and V. For the estimation of θ, given β and V, the complete data log-likelihood is given by

$$\ell(\beta,\sigma^2,D) = \sum_{i=1}^{N}\log\left[f(y_i \mid \beta,\sigma^2,b_i)\right] + \sum_{i=1}^{N}\log\left[f(b_i \mid D)\right], \qquad (8.47)$$

which, apart from a constant can be expressed as

$$\ell(\beta,\sigma^2,D) = -\sum_{i=1}^{N}\left[\frac{n_i\log\sigma^2}{2} + \frac{1}{2\sigma^2}(y_i - X_i\beta - Z_ib_i)^{\mathrm{T}}(y_i - X_i\beta - Z_ib_i)\right]$$
$$- \sum_{i=1}^{N}\left[\frac{q\log(2\pi)}{2} + \frac{1}{2}\log|D| + \frac{1}{2}b_i^{\mathrm{T}}D^{-1}b_i\right]. \qquad (8.48)$$

This expression establishes that $\sum_{i=1}^{N}(y_i - X_i\beta - Z_ib_i)^{\mathrm{T}}(y_i - X_i\beta - Z_ib_i) \equiv \sum_{i=1}^{N}e_i^{\mathrm{T}}e_i$ and $\sum_{i=1}^{N}b_i^{\mathrm{T}}b_i$ are the complete data sufficient statistics for σ^2 and D, respectively. Then it can be seen that

$$\hat{\sigma}^2 = \frac{\sum_{i=1}^{N}e_i^{\mathrm{T}}e_i}{\sum_{i=1}^{N}n_i}, \quad \text{and} \qquad \hat{D} = \sum_{i=1}^{N}\frac{b_i^{\mathrm{T}}b_i}{N}. \qquad (8.49)$$

This completes the M-step. In the E-step we calculate the expected value of the sufficient statistics given the observed data and the current parameter estimates as

$$\mathbf{E}(b_ib_i^{\mathrm{T}} \mid y_i,\hat{\beta},\hat{\sigma}^2,\hat{D}) = \mathbf{E}(b_i \mid y_i,\hat{\beta},\hat{\sigma}^2,\hat{D})\mathbf{E}(b_i^{\mathrm{T}} \mid y_i,\hat{\beta},\hat{\sigma}^2,\hat{D}) + \mathrm{Var}(b_i \mid y_i,\hat{\beta},\hat{\sigma}^2,\hat{D})$$
$$= \hat{D}Z_i^{\mathrm{T}}\hat{\Sigma}_i^{-1}(y_i - X_i\hat{\beta})(y_i - X_i\hat{\beta})^{\mathrm{T}}\hat{\Sigma}_i^{-1}Z_i\hat{D} + \hat{D} - \hat{D}Z_i^{\mathrm{T}}\hat{\Sigma}_i^{-1}Z_i\hat{D}$$

and

$$\mathbf{E}(e_ie_i^{\mathrm{T}} \mid y_i,\hat{\beta},\hat{\sigma}^2,\hat{D}) = \mathrm{tr}(\mathbf{E}(e_ie_i^{\mathrm{T}} \mid y_i,\hat{\beta},\hat{\sigma}^2,\hat{D}))$$
$$= \mathbf{E}(e_i \mid y_i,\hat{\beta},\hat{\sigma}^2,\hat{D})\mathbf{E}(e_i^{\mathrm{T}} \mid y_i,\hat{\beta},\hat{\sigma}^2,\hat{D}) + \mathrm{Var}(e_i \mid y_i,\hat{\beta},\hat{\sigma}^2,\hat{D})$$
$$= \mathrm{tr}(\hat{\sigma}^4\hat{\Sigma}_i^{-1}(y_i - X_i\hat{\beta})(y_i - X_i\hat{\beta})^{\mathrm{T}}\hat{\Sigma}_i^{-1} + \sigma^2 I_{n_i} - \hat{\sigma}^4\hat{\Sigma}_i^{-1}),$$

where $\hat{\Sigma}_i = Z_i \hat{D} Z_i^T + \hat{\sigma}^2 I_{n_i}$ and $e_i = y_i - X_i \beta - Z_i b_i$. Then, the maximum likelihood estimates of all the parameters are obtained as

Step 1: Given some initial estimates $\hat{\sigma}^2$ and $\hat{D}$ of σ^2 and D. Obtain $\hat{\Sigma}_i = Z_i \hat{D} Z_i^T + \hat{\sigma}^2 I_{n_i}$. Use these to estimate β as

$$\hat{\beta} = \left(\sum_{i=1}^{N} X_i^T \Sigma_i^{-1} X_i \right)^{-1} \sum_{i=1}^{N} X_i^T \Sigma_i^{-1} y_i. \tag{8.50}$$

Step 2: Use $\hat{\sigma}^2$, $\hat{D}$, $\hat{\Sigma}_i$, and $\hat{\beta}$ from step 1 to obtain

$$\hat{\sigma}^2 = \frac{\sum_{i=1}^{N} e_i^T e_i}{\sum_{i=1}^{N} n_i} \tag{8.51}$$

and

$$\hat{D} = \sum_{i=1}^{N} \frac{b_i^T b_i}{N}, \tag{8.52}$$

where the ith component in the numerator on the right hand side of equations (8.51) and (8.52) are $\mathrm{tr}(\hat{\sigma}^4 \hat{\Sigma}_i^{-1} (y_i - X_i \hat{\beta})(y_i - X_i \hat{\beta})^T \hat{\Sigma}_i^{-1} + \sigma^2 I_{n_i} - \hat{\sigma}^4 \hat{\Sigma}_i^{-1})$ and $\hat{D} Z_i^T \hat{\Sigma}_i^{-1} (y_i - X_i \hat{\beta})(y_i - X_i \hat{\beta})^T \hat{\Sigma}_i^{-1} Z_i \hat{D} + \hat{D} - \hat{D} Z_i^T \hat{\Sigma}_i^{-1} Z_i \hat{D}$, respectively. Finally, the maximum likelihood estimates of the parameters β, σ^2 and D are obtained by iterating between step 1 and step 2. For a discussion on convergence issues of the EM algorithm here see Ibrahim and Molenberghs (2009).

Example 4. The six cities study of air pollution and health was a longitudinal study designed to characterize lung growth as measured by changes in pulmonary function in children and adolescents, and the factors that influence lung function growth. A cohort of 13,379 children born on or after 1967 was enrolled in six communities across the U.S.: Watertown (Massachusetts), Kingston and Harriman (Tennessee), a section of St. Louis (Missouri), Steubenville (Ohio), Portage (Wisconsin), and Topeka (Kansas). Most children were enrolled in the first or second grade (between the ages of six and seven) and measurements of study participants were obtained annually until graduation from high school or loss to follow-up. At each annual examination, spirometry, the measurement of pulmonary function, was performed and a respiratory health questionnaire was completed by a parent or guardian. The basic maneuver in simple spirometry is maximal inspiration (or breathing in) followed by forced exhalation as rapidly as possible into a closed container. Many different measures can be derived from the spirometric curve of volume exhaled versus time. One widely used measure is the total volume of air exhaled in the first second of the maneuver (FEV_1).

Fitzmaurice et al. (2004) present an analysis of a subset of the pulmonary function data collected in the Six Cities Study. The dataset contains a subset of the pulmonary function data collected in the Six Cities Study. The data consist of all measurements of FEV1, height and age obtained from a randomly selected subset of the female participants living in Topeka, Kansas. The random sample consists of 300 girls, with a minimum of one and a maximum of twelve observations over time. Data for four selected girls are presented in Table 8.1 in Fitzmaurice et al. (2004). These data were then analysed using the following regression model

$$Y_{ij} = \beta_1 + \beta_2 Age_{ij} + \beta_3 \log(height_{ij}) + \beta_3 Age_{i1} + \beta_4 \log(height_{i1}) + b_{1i} + b_{2i} Age_{ij}. \tag{8.53}$$

The estimates of the regression coefficients (fixed effect) and standard errors for the $\log(FEV_1)$ are given in their Table 8.2 (page 213).

Exercise 5.

Obtain maximum likelihood estimates of the parameters of the model in page 213 for the Six Cities Study of Air Pollution and Health data given in page 211 of Fitzmaurice et al. (2004) by the EM algorithm described above using R.

8.6.3 Estimation with Nonignorable Missing Response Data (MAR and MNAR)

The method of estimation used for complete data will now be extended to missing data under MAR and MNAR mechanism used in equations (8.1) and (8.1) in Section 8.2. In MAR, conditional probability of missingness of the data depends on the observed data. Parameters of the missingness mechanism are completely separate and distinct from the parameters of the model (8.39). In likelihood based estimation considering MAR, missingness mechanism can be ignored from the likelihood and missing data that are missing at random are often known as ignorable missing or ignorable nonresponse, but the subjects having these missing observations cannot be deleted before the analysis. That is, in MAR the distribution of r_i given in equation (8.1) is not required but in MNAR we need both models given in (8.1) and (8.2). So, in what follows we deal with the analysis under MNAR mechanism. Because the part of the likelihood involving r_i does not depend on the parameters of interest; therefore, the MLE of the parameters of interest is the same without the part involving r_is. So for MAR the distribution of r_i is deleted.

To put things in perspective, define a $n_i \times 1$ random vector R_i, whose jth component has the binary distribution

$$r_{ij} = \begin{cases} 0 & \text{if } y_{ij} \text{ is observed,} \\ 1 & \text{if } y_{ij} \text{ is missing.} \end{cases} \tag{8.54}$$

The distribution of R_i follows a multinomial distribution with 2^{n_i} cell probabilities indexed by the parameter vector ϕ. We assume that the covariates are fully observed. Under the normal mixed model, the complete data density of (y_i, b_i, r_i) for subject i is then given by $f(y_i, b_i, r_i \mid \beta, \sigma^2, D, \phi)$. Little (1993, 1995) suggested that this can be factored as

$$f(y_i, b_i, r_i \mid \beta, \sigma^2, D, \phi) = f(y_i \mid \beta, \sigma^2, b_i) f(b_i \mid D) f(r_i \mid \phi, y_i) \tag{8.55}$$

(the selection model) or

$$f(y_i, b_i, r_i \mid \beta, \sigma^2, D, \phi) = f(y_i \mid \beta, \sigma^2, b_i, r_i) f(b_i \mid D) f(r_i \mid \phi) \tag{8.56}$$

(the pattern mixture model). Here we consider only the selection model. Now, let $\gamma = (\beta, \sigma^2, D, \phi)$. Then, the complete data log-likelihood is

$$
\ell(\gamma) = \log\left[\prod_{i=1}^{N} f(y_i, b_i, r_i \mid \beta, \sigma^2, D, \phi)\right]
$$

$$
= \sum_{i=1}^{N} \log f(y_i \mid \beta, \sigma^2, b_i) + \sum_{i=1}^{N} \log f(b_i \mid D) + \sum_{i=1}^{N} \log f(r_i \mid \phi, y_i). \tag{8.57}
$$

The parameters β, σ^2, and D are usually of interest, while considering b_i and ϕ as nuisance parameters. Diggle and Kenward (1994) discussed estimation methods assuming monotone missing data. However, as Ibrahim and Molenberghs (2009) comment, these methods are not easily extended to the analysis of nonmonotone missing data, where a subject may be observed after a missing value occurs. A weighted EM method, proposed by Ibrahim (1990), which applies to both monotone and nonmonotone missing data, is given below.

Writing $y_i = (y_{mis,i}, y_{obs,i})$, where $y_{mis,i}$ is the $s_i \times 1$ vector of missing components of y_i, the E-step calculates the expected value of the complete data log-likelihood given the observed data and current parameter estimates as

$$
Q_i(\gamma \mid \gamma^{(t)}) = \mathbf{E}(l(\gamma; y_i, b_i, r_i) \mid y_{obs,i}, r_i, \gamma^{(t)}), \tag{8.58}
$$

where $\gamma^{(t)} = (\beta^{(t)}, \sigma^{2(t)}, D^{(t)}, \phi^{(t)})$. Since both b_i and $y_{mis,i}$ are unobserved, they must be integrated out. Thus, the E-step for the ith observation at the $(t+1)$th iteration is

$$
Q_i(\gamma \mid \gamma^{(t)}) = \int \int \log[f(y_i \mid \beta, \sigma^2, b_i)] f(y_{mis,i}, b_i \mid y_{obs,i}, r_i, \gamma^t) db_i dy_{mis,i}
$$

$$
+ \int \int \log[f(b_i \mid D)] f(y_{mis,i}, b_i \mid y_{obs,i}, r_i, \gamma^t) db_i dy_{mis,i}
$$

$$
+ \int \int \log[f(r_i \mid \phi, y_i)] f(y_{mis,i}, b_i \mid y_{obs,i}, r_i, \gamma^t) db_i dy_{mis,i}
$$

$$
\equiv I_1 + I_2 + I_3, \tag{8.59}
$$

where $f(y_{mis,i}, b_i \mid y_{obs,i}, r_i, \gamma^t)$ represents the conditional distribution of the data considered (missing), given the observed data.

Detailed calculations, which are omitted here, given in Ibrahim and Molenberghs (2009), lead to the evaluation of $Q_i(\gamma \mid \gamma^{(t)})$ as given below. Let $u_{i1}, \ldots, u_{im_i}$ be a sample of size m_i from

$$
f(y_{mis,i} \mid y_{obs,i}, r_i, \gamma^{(t)}) \propto \exp\left\{-\frac{1}{2}(y_i - X_i\beta^{(t)})^{\mathrm{T}}(Z_i D^{(t)} Z_i^{\mathrm{T}} + \sigma^{2(t)} I_{n_i})^{-1}(y_i - X_i\beta^{(t)})\right\}
$$

$$
\times f(r_i \mid y_{mis,i}, y_{obs,i}, \gamma^{(t)}), \tag{8.60}
$$

obtained via the Gibbs sampler along with the adaptive rejection algorithm of Gilks and Wild (1992), where the specific model for $f(r_i \mid y_{mis,i}, y_{obs,i}, \gamma^{(t)})$ is given in their equation (5.43). Also see Ibrahim and Molenberghs (2009) for further discussion.

Now let $y_i^{(k)} = (u_{ik}^T, y_{obs,i}^T)^T$ and $b_i^{(tk)} = \Sigma_i^{(t)} Z_i^T (y_i^{(k)} - X_i \beta^{(t)})/\sigma^{2(t)}$. Then, the E-step for the ith observation at the $(t+1)$th iteration can be simplified as

$$
Q_i(\gamma|\gamma^{(t)}) = -\frac{n_i}{2}\log(2\pi) - \frac{n_i \log(\sigma^2)}{2} - \frac{1}{2\sigma^2} tr(Z_i^T Z_i \Sigma_i^{(t)})
$$

$$
+ \frac{1}{m_i} \sum_{k=1}^{m_i} (y_i^{(k)} - X_i\beta - Z_i b_i^{(tk)})^T (y_i^{(k)} - X_i\beta - Z_i b_i^{(tk)})
$$

$$
- \frac{q\log(2\pi)}{2} - \frac{\log(|D|)}{2} - \frac{1}{2} tr(D^{-1}\Sigma_i^{(t)})
$$

$$
- \frac{1}{2m_i} \sum_{k=1}^{m_i} [(b_i^{(tk)})^T D^{-1} b_i^{(tk)}] + \frac{1}{m_i} \sum_{k=1}^{m_i} \log[f(r_i|\phi, y_i^{(k)})] \qquad (8.61)
$$

and hence for all N observations it is

$$
Q(\gamma|\gamma^{(t)}) = \sum_{i=1}^{N} Q_i(\gamma|\gamma^{(t)}). \qquad (8.62)
$$

In the M-step $Q(\gamma|\gamma^{(t)})$ is to be maximized to obtain estimates of β, σ^2, and D. It can be shown that $\phi^{(t+1)}$ is obtained by maximizing

$$
Q_\phi = \sum_{i=1}^{N} \frac{1}{m_i} \sum_{k=1}^{m_i} \log[f(r_i|\phi, y_i^k)] \qquad (8.63)
$$

and that closed form estimates for β, σ^2, and D are

$$
\beta^{(t+1)} = \Big(\sum_{i=1}^{N} X_i^T X_i\Big)^T \sum_{i=1}^{N} \Big(\frac{1}{m_i} \sum_{k=1}^{m_i} X_i^T (y_i^{(k)} - Z_i b_i^{(tk)})\Big), \qquad (8.64)
$$

$$
\sigma^{2(t+1)}
$$

$$
= \frac{1}{M} \sum_{i=1}^{N} \left[\frac{1}{m_i} \sum_{k=1}^{m_i} (y_i^{(k)} - X_i\beta^{(t+1)} - Z_i b_i^{(tk)})^T (y_i^{(k)} - X_i\beta^{(t+1)} - Z_i b_i^{(tk)}) + tr(Z_i^T Z_i \Sigma_i^{(t)}) \right]
$$

$$
(8.65)
$$

and

$$
D^{(t+1)} = \frac{1}{N} \sum_{i=1}^{N} \Big[\frac{1}{m_i} \sum_{k=1}^{m_i} b_i^{(tk)} (b_i^{(tk)})^T + \Sigma_i^{(t)} \Big], \qquad (8.66)
$$

where $M = \sum_{i=1}^{N} n_i$. MLEs of β, σ^2, and D are obtained by iterating between the E-step and the M-step.

Exercise 6. Delete at random 10% of the response data from the data given in Example 4. For these new data, obtain maximum likelihood estimates of the parameters of the model in page 213 for the data given in page 211 of Fitzmaurice et al. (2004) using R.

In biostatistics or similar fields many longitudinal studies involve nonignorable missing data. Here we present two data sets.

Example 5. (IBCSG data)
Consider a data set concerning the quality of life among breast cancer patients in a clinical trial comparing four different chemotherapy regimens conducted by the International Breast Cancer Study Group (IBCSG Trial VI; Ibrahim et al. 2001). The main outcomes of the trial were time until relapse and death, but patients were also asked to complete quality of life questionnaires at baseline and at three-month intervals. Some patients did refuse, on occasion, to complete the questionnaire. However, even when they refused, the patients were asked to complete an assessment at their next follow-up visit. Thus, the structure of this trial resulted in nonmonotone patterns of missing data. One longitudinal quality of life outcome was the patient's self-assessment of her mood, measured on a continuous scale from 0 (best) to 100 (worst). The three covariates of interest included a dichotomous covariate for language (Italian or Swedish), a continuous covariate for age, and three dichotomous covariates for the treatment regimen (4 regimens). Data from the first 18 months of the study were used, implying that each questionnaire was filled out at most seven times, i.e., at baseline plus at six follow-up visits.

There are 397 observations in the data set, and mood is missing at least one time for 71% of the cases, resulting in 116 (29%) complete cases. The amount of missing data is minimal at baseline (2%) and ranges between 24% and 31% at the other six times: 26.2% at the second, 24.2% at the third, 29% at the fourth, 24.9% at the fifth, 28.2% at the sixth, and 30.5% at the seventh occasion. For details of the data and analysis, see Ibrahim et al. (2001). All patients were alive at the end of 18 months, so no missingness is due to death. However, it is reasonable to conjecture that the mood of the patient affected their decision to fill out the questionnaire. In this case, the missingness would be MNAR, and an analysis that does not include the missing data mechanism would be biased.

Example 6. (Muscatine children's obesity data)
The Muscatine Coronary Risk Factor Study (MCRFS) was a longitudinal study of coronary risk factors in school children (Woolson and Clarke 1984; Ekholm and Skinner 1998). Five cohorts of children were measured for height and weight in 1977, 1979, and 1981. Relative weight was calculated as the ratio of a child's observed weight to the median weight for their age-sex-height group. Children with a relative weight greater than 110% of the median weight for their respective stratum were classified as obese. The analysis of this study involves binary data (1 = obese, 0 = not obese) collected at successive time points. For every cohort, each of the following seven response patterns occurs: (p,p,p), (p,p,m), (p,m,p), (m,p,p), (p,m,m), (m,p,m), and (m,m,p), where a p (m) denotes that the child was present (missing) for the corresponding measurement. The distribution over the patterns is shown in Table 1 of Ekholm and Skinner (1998).

The statistical problem is to estimate the obesity rate as a function of age and sex. However, as can be seen in Table 1 of Ekholm and Skinner 1998) , many data records are incomplete since many children participated in only one or two occasions of the survey. Ekholm and Skinner (1998) report that the two main reasons for nonresponse were: (i) no consent form signed by the parents was received, and (ii) the child was not in school on the day of the examination. If the parent did not sign the consent form because they did not want their child to be labeled as obese, or if the child did not attend school the day of the survey because of their weight, then the missingness would at least be MAR, and likely even MNAR. In the latter case, an analysis that ignores the missing data mechanism would be biased. However, since the outcome is binary, these data cannot be modeled using the normal random effects model. Instead, a general method based on the generalized estimating equations would be useful (see Section 8.6.4).

8.6.4   Generalized Estimating Equations

8.6.4.1   Introduction

In Sections 8.6.2 and 8.6.3, we show analysis of missing data when the data in question follow a particular model. However, in practice model departure is often a challenging issue. For example, when there is evidence of over-dispersion in count data, a negative binomial distribution is used. Similarly, when data in the form of proportions arise, a binomial model may not be the correct model as over-dispersion may be present in the data. In this case, often a Beta-binomial distribution is used. See, for example, Paul and Plackett (1978) and Paul and Islam (1995). In all these data analysis situations, a particular model for the data is assumed. But, in practice, data may follow a different distribution than what is assumed. For example, over-dispersed count data may have come from a different over-dispersed count data model than a negative binomial model, such as a log-normal mixture of the Poisson distribution. Theoretical analysis with a log-normal mixture of the Poisson distribution is complex at best. Yet this assumption may not be valid. So, robust data analytic methods, such as, the quasi-likelihood (Wedderburn, 1974) and the extended quasi-likelihood (Nelder and Pregibon, 1987; Goodambe and Thompson, 1989), the double extended quasi-likelihood (Lee and Nelder, 2001) have been developed. However, these methods are for data in which distribution of observations can be assumed to be independent. For the analysis of longitudinal data, a robust procedure using generalized estimating equations (GEE) has been developed by Liang and Zeger (1986) and Zeger and Liang (1986).

The GEE approach of obtaining estimates of the regression parameters requires that there is no missing data, that is when the missingness probability does not depend on the responses. No general theory is available for GEE estimates of the regression parameters in presence of missing data. However, there are a few very useful special cases which are given below.

8.6.4.2   Weighted GEE for MAR Data

Robins et al. (1995) proposed a weighted GEE method for analysis for data that are MAR. Let y_{it} and x_{it} be the response and a vector of independent variables for individual i at time t, $t = 1, \ldots, n_i$, where n_i denotes the number of repeated observations for the ith subject. Denote the complete longitudinal outcomes and independent variables by $Y_i = (y_{i1}, \ldots, y_{in_i})^T$ and $X_i = (x_{i1}, \ldots, x_{in_i})^T$ and let $E(Y_i|x_i) = \mu_i = (\mu_{i1}, \mu_{i2}, \ldots, \mu_{in_i})^T$. For completely observed longitudinal data, the GEE approach of Liang and Zeger (1986) allowed regression modeling of the data specifying only the mean and variance of the outcome variables. The mean for the complete data is assumed to satisfy $\mu_{it} = g^{-1}(\eta_{it}) = g^{-1}(x_{it}\beta)$, where $g(\cdot)$ is a link function and the estimating equation for β is

$$U_1(\beta, \alpha) = \sum_{i=1}^{N} \left(\frac{\partial \mu_i}{\partial \beta} \right)^T V_i^{-1}(\beta, \alpha)(Y_i - \mu_i) = 0, \tag{8.67}$$

where $V_i(\beta, \alpha) = \phi A_i(\mu_i)^{1/2} R_i(\rho) A_i(\mu_i)^{1/2}$, $A_i = diag\{var(y_{i1}), \ldots, var(y_{in_i})\}$, $\alpha = (\phi, \rho)$, and $R_i(\rho)$ is a working correlation matrix of Y_i. For the choice of $R_i(\rho)$, see Chapter 4. Denote the GEE estimate of β obtained by solving $U_1(\beta, \hat{\alpha}) = 0$ as $\hat{\beta}_1$, where $\hat{\alpha}$ is any $\sqrt{N}$-consistent estimator. Usually, $\hat{\alpha}$ is estimated by the method of moments Zhang and Paul (2013). Liang and Zeger (1986) showed that $\hat{\beta}_1$ is consistent and asymptotically normal and its variance can be consistently estimated by a sandwich-type estimator (see Chapter 4).

As earlier, let r_{it} be the missing data indicator variable for the tth observation of the ith individual and assume monotone missing with $r_{i1} \geq r_{i2} \geq \ldots r_{iN-1} \geq r_{iN}$ (that is, an individual who leaves the study never comes back). Also, assume, $r_{i1} = 1$ for all i; that is, all responses are completely observed at baseline.

Then, the weighted estimating equation (Robins and Rotnitzky 1995, Preisser, Lohman, and Rathouz 2002) for obtaining unbiased GEE estimates of the regression parameters under MAR is

$$U_2(\beta, \alpha, \gamma) = \sum_{i=1}^{N} \left(\frac{\partial \mu_i}{\partial \beta} \right)^T V_i^{-1}(\beta, \alpha) \Delta_i (Y_i - \mu_i) = 0, \tag{8.68}$$

where the jth diagonal element of Δ_i is $r_{ij} w_{ij}$, $w_{ij} = 1/P(r_{ij} = 1|X_i, Y_i)$, and γ is a parameter from the distribution of r_{ij}.

The weights w_{ij} are generally estimated using a logistic regression model under MAR assumption. See Lin and Rodrigues (2015) for further details.

Robins et al. (1995) showed that if Δ_i is estimated consistently, then $\hat{\beta}_2$ is consistent and asymptotically normal under MAR and monotone missing patterns.

8.6.5   Some Applications of the Weighted GEE

8.6.5.1   Weighted GEE for Binary Data

Following Robins et al. (1995), Troxel et al. (1997) developed a weighted GEE method for nonignorably missing binary response survey data. Let $Y = (Y_1, \ldots, Y_N)$ represent the underlying response (zero and one) vector for the N independent subjects and let X_i be the vector of p covariates of the ith subject. Suppose that the initial nonresponders are followed up and given a second opportunity to respond to the survey. The indicator variable $r_1 = (r_{i1}, \ldots, r_{n1})$ indicates if (Y_i, X_i) is observed or not at the first time, and similarly, r_2 is the indicator variable for the second time for those who did not respond at the first time. The missingness probabilities are assumed to depend on the unobserved response variable Y_i and a set of the co-variates X_i^*. Let $\pi_{i1} = P(r_{i1} = 1 | Y_i, X_i^*)$ and $\pi_{i2} = P(r_{i2} = 1 | Y_i, X_i^*, r_{i1} = 0)$. Suppose $\mu_i = \mathbf{E}(Y_i | X_i) = g(X_i\beta)$ and that β are parameters of interest. Based on the available data from subjects who responded either at the first time or at the second time, the parameter β can be estimated by solving the estimating equations

$$\sum_{i=1}^{N} r_i E_i^T W_i^{-1}(Y_i - \mu_i) = 0, \tag{8.69}$$

where $E_i = \partial\mu_i/\partial\beta$, W_i is the variance of Y_i, and $r_i = r_{i1} + r_{i2}$.

As pointed out by Troxel et al. (1997), if r_{i1} or r_{i2} depends on Y_i, the above estimating equations are biased from zero and hence result in biased estimates. To this end, Troxel et al. (1997) proposed to use the weighted estimating functions

$$U^{TLB} = \sum_{i=1}^{N} \frac{r_i}{\theta_i} E_i^T W_i^{-1}(Y_i - \mu_i), \tag{8.70}$$

where $E_i = \partial\mu_i/\partial\beta$, W_i is the variance of Y_i, $r_i = r_{i1} + r_{i2}$, and $\theta_i = \pi_{i1} + (1 - \pi_{i1})\pi_{i2}$ is the probability of being observed. It can be shown that $\mathbf{E}(U^{TLB}) = 0$, and hence $U^{TLB} = 0$ will produce consistent estimates of β under very general conditions.

8.6.5.2 Two Modifications

Wang (1999b) modified the estimating functions suggested by Robins et al. (1995) and suggested two alternative modifications for unbiased estimation of regression parameters when a binary outcome is potentially observed at successive time points. In particular, he proposed two new ways of constructing estimating functions, one of which is based on direct bias correction and the other is motivated by condition-ing. Unlike the weighting methods, the two estimating functions proposed by Wang (1999b) are score-like functions and are, therefore, the most efficient. Also, they are unbiased even for nonignorably missing data.

Modified Approach of Robins et al. (1995)

A simple modification of the approach by Robins et al. (1995) leads to the fol-lowing unbiased estimating functions:

$$U^{RRZ} = \sum_{i=1}^{N} \frac{r_{i1}}{\pi_{i1}} E_i^T W_i^{-1}(Y_i - \mu_i)$$

$$+ \sum_{i=1}^{N} \frac{r_{i2}}{(1 - \pi_{i1})\pi_{i2}} E_i^T W_i^{-1}(Y_i - \mu_i). \tag{8.71}$$

The weight here depends on when the response is observed (at time 1 or time 2), while U^{TLB} uses the overall probability of being observed either at time 1 or time 2.

Bias-Corrected Estimating Functions

The weighting GEE approach of Troxel et al. (1997) requires that (a) an individual will be further observed even if he or she has been observed earlier and (b) return is not possible once a subject leaves the study. Both of these conditions can be avoided if we use $U^{CC} = \mathbf{E}(U^{CC})$ rather than $U^{CC} = 0$ for parameter estimation. Wang (1999b) showed that we can write the unbiased estimating equations $U^{(1)} = U^{CC} - \mathbf{E}(U^{CC})$ as

$$U^{(1)} = \sum_{i=1}^{N} r_i E_i^T W_i^{-1} (Y_i - \eta_i), \qquad (8.72)$$

where $\eta_i = P(Y_i|X_i, R_i = 1)$. This can be also be expressed as a function of μ_i and $\eta_i = \theta_{i1}\mu_i/\theta_{i1}\mu_i + (1-\mu_i)\theta_{i0}$ as

$$U^{(1)} = \sum_{i=1}^{N} r_i \left(\frac{\partial \eta_i}{\partial \beta}\right)^T \frac{Y_i - \eta_i}{\text{Var}(Y_i|R_i = 1)}.$$

This implies that $U^{(1)}$ may be derived by maximizing $\prod_{R_i=1} P(Y_i|r_i = 1)$. The form of $U^{(1)}$ also suggests that they are the most efficient estimating functions (conditional on $r_i = 1$) in terms of asymptotic variance of the estimates (Godambe and Heyde, 1987).

Conditional Estimating Functions

The complete-case model produces biased estimates because the expectations of Y_i for those who responded differ from μ_i. The estimating functions given by (8.69) and (8.70) are correct for the biases due to the response-dependent selection. However, responses obtained at time 1 are not distinguished from those obtained at time 2 in U^{TLB}.

Let $\mu_{i1} = \mathbf{E}(Y_i|X_i, r_{i1} = 1)$ and $\mu_{i2} = \mathbf{E}(Y_i|X_i, r_{i1} = 0, r_{i2} = 1)$. Parameter estimation can rely on the following unbiased estimating functions:

$$U^{(2)} = \sum_{i=1}^{N} r_{i1} D_{i1}^T W_{i1}^{-1} (Y_i - \mu_{i1}) + \sum_{i=1}^{N} r_{i2} D_{i2}^T W_{i2}^{-1} (Y_i - \mu_{i2}), \qquad (8.73)$$

in which $D_{ik} = \partial \mu_{ik}/\partial \beta$ and $W_{ik} = \mu_{ik}(1 - \mu_{ik})$ for $k = 1$ and 2.

We will write the missingness probabilities as $\pi_{i1}(Y_i)$ and $\pi_{i2}(Y_i)$ to express the dependence on the response variable Y_i explicitly. The relationships between μ_{i1}, μ_{i2} and μ_1 are given by

$$\mu_{i1} = \frac{\pi_{i1}(1)\mu_i}{\pi_{i1}(1)\mu_i + \pi_{i1}(0)(1 - \mu_i)} \qquad (8.74)$$

and

$$\mu_{i2} = \frac{\pi_{i2}(1)\{1 - \pi_{i1}(1)\}\mu_i}{\pi_{i2}(1)\{1 - \pi_{i1}(1)\}\mu_i + \pi_{i2}(0)\{1 - \pi_{i1}(0)\}(1 - \mu_i)}. \qquad (8.75)$$

Because $D_{ik}^T W_{ik}^{-1} = E_i^T W_i^{-1}$, (5.67) can be re written as

$$U^{(2)} = \sum_{i=1}^{N} \left[E_i^T W_i^{-1} \{ r_{i1}(Y_i - \mu_{i1}) + r_{i2}(Y_i - \mu_{i2}) \} \right]. \qquad (8.76)$$

Asymptotic Covariance

Wang (1999b) discussed the estimation of the asymptotic covariance of the GEE estimates of the regression parameters β. Omitting details, the asymptotic covariance of the estimates, $\hat{\beta}$, obtained from the unbiased estimating equations $U = 0$ discussed above, can be approximated by $V = A^{-1}B(A^T)^{-1}$, where $A = \mathbf{E}(\partial U/\partial\beta)$ and $B = \mathbf{E}(UU^T)$, the covariance matrix of the estimating functions. Table 1 of Wang (1999b), gives asymptotic covariance of the estimates.

Example 7.

Wang (1999b) analyzed a set of data given in Troxel et al. (1997) from a survey study concerning medical practice guidelines. The purpose of the survey was to assess the lawyers' knowledge of medical practice guidelines and the effect of those guidelines on their legal work. Overall, 578 responses were received (among the 960 recipients) from the two mailing waves. Because of missing values in the received responses, there are only 524 subjects with complete information on four binary variables. The four binary variables are Years (= 1 if > 10 years of law practice), Firm P (= 1 if > 20% of the firm's work has involved malpractice), Number (= 1 if > 5 new cases a year), and the response variable, Aware (awareness of medical practice guidelines). Wang (1999) used the same model as in Troxel et al. (1997). To eliminate the complications arising from the missing values in the responses received, he only used observations from the 524 subjects and adjusted the total number of recipients to 960(524)/578.

The variables used in the missingness model are Years, Firm P, Number, Aware, Years × Aware, and Firm P × Number. For these variables, a logit regression model was used which also included an intercept parameter and a parameter τ representing amounts of missingness. The parameters in the missingness model were estimated as $\alpha = (-1.75, 0.72, 1.11, 1.65, 1.12, -1.14, -1.85, 0.15)$ with standard errors $(0.35, 0.65, 0.67, 0.50, 0.60, 0.79, 0.78, 0.15)$. The parameter estimates of the four estimating function methods are quite different from those of the CC analysis. For the intercept, Years, Firm p, and Number, the coefficient estimates from the CC analysis are $(-0.19, 0.01, 0.25, 0.62)$. The corresponding estimates from the weighting methods of TLB and RRZ are $(-0.75, 0.49, 0.30, 0.71)$ with standard errors $(0.35, 0.42, 0.20, 0.22)$. Estimates from the new methods $U^{(1)}$ and $U^{(2)}$ are similar to those from the weighting methods except for one variable (Years). The effects of Years estimated by $U^{(1)}$ and $U^{(2)}$ are close to zero, quite different from the TLB and RRZ estimates. Considering the large standard errors for the Year effect, we should not be surprised by the difference. Also notice that the missingness model plays an important role in estimating β, and the effect of Year in the missingness model is also very uncertain, which may have a serious impact on the estimate of the Year effect.

Estimation with ignorable missing response data

For estimation with ignorable missing response data the missingness mechanism can be ignored. So the joint distribution would be $f(y_i, b_i \mid \beta, \sigma^2, D)$ instead of $f(y_i, b_i, r_i \mid \beta, \sigma^2, D, \phi)$ in case of nonignorable missing response data. For ignorable (MAR) missing response data, selection models decompose the joint distribution as

$$f(y_i, b_i \mid \beta, \sigma^2, D) = f(y_i \mid \beta, \sigma^2, b_i)f(b_i \mid D). \tag{8.77}$$

The rest of the estimation procedure would remain similar to nonignorable missing response data.

Using the same framework, methods for analysis of longitudinal (or otherwise) data with missing values can be developed when data follow other continuous or discrete models. For example, Ibrahim and Lipsitz (1996) developed estimation procedures for parameters of a Binomial Regression in presence of nonignorable missing responses.

Chapter 9

Random Effects and Transitional Models

9.1 A General Discussion

Traditional designed experiments involve data that need to be analyzed using a fixed effects model or a random effects model. Central to the idea of variance components models is the idea of fixed and random effects. Each effect in a variance components model must be classified as either a fixed or a random effect. Fixed effects arise when the levels of an effect constitute the entire population about which one is interested. For example, if a plant scientist is comparing the yields of three varieties of soybeans, then Variety would be a fixed effect, provided that the scientist is concerned about making inferences on only these three varieties of soybeans. Similarly, if an industrial experiment focused on the effectiveness of two brands of a machine, machine would be a fixed effect only if the experimenter's interest does not go beyond the two machine brands.

On the other hand, an effect is classified as a random effect when one wants to make inferences on an entire population, and the levels in your experiment represent only a sample from that population. Psychologists comparing test results between different groups of subjects would consider Subject as a random effect. Depending on the psychologists' particular interest, the group effect might be either fixed or random. For example, if the groups are based on the sex of the subject, then sex would be a fixed effect. But if the psychologists are interested in the variability in test scores due to different teachers, then they might choose a random sample of teachers as being representative of the total population of teachers, and teacher would be a random effect. Note that, in the soybean example presented earlier, if the scientists are interested in making inferences on the entire population of soybean varieties and randomly choose three varieties for testing, then variety would be a random effect.

If all the effects in a model (except for the intercept) are considered random effects, then the model is called a random effects model; likewise, a model with only fixed effects is called a fixed-effects model. The more common case, where some factors are fixed and others are random, is called a mixed model.

In the linear model, each level of a fixed effect contributes a fixed amount to the expected value of the dependent variable. What makes a random effect different is that each level of a random effect contributes an amount that is viewed as a sample

DOI: 10.1201/9781315153636-9

from a population of normally distributed variables, each with mean 0, and an unknown variance, much like the usual random error term that is a part of all linear models. The variance associated with the random effect is known as the variance component because it is measuring the part of the overall variance contributed by that effect.

9.2   Random Intercept Models

Suppose m large elementary schools are chosen randomly from among thousands in a large country. Suppose also that n pupils of the same age are chosen randomly at each selected school. Their scores on a standard aptitude test are ascertained. Let Y_{ij} be the score of the jth pupil at the ith school. A simple way to model the relationships of these quantities is

$$Y_{ij} = \mu + \alpha_i + \eta_{ij},$$

where μ is the average test score for the entire population. In this model α_i is the school-specific random effect: it measures the difference between the average score at school i and the average score in the entire country. The term η_{ij} is the individual-specific effect, i.e., it's the deviation of the jth pupil score from the average for the ith school.

The model can be augmented by including additional explanatory variables, which would capture differences in scores among different groups. For example:

$$Y_{ij} = \mu + \beta_1 \text{Sex}_{ij} + \beta_2 \text{Race}_{ij} + \beta_3 \text{ Parentedu}_{ij} + \alpha_i + \eta_{ij},$$

where Sex_{ij} is the dummy variable for boys/girls, Race_{ij} is the dummy variable for white/black pupils, and Parenteduc_{ij} records the average education level of child's parents. This is a mixed model, not a purely random effects model, as it introduces fixed-effects terms for Sex, Race, and Parents' Education. A simple one-way random effects model was given in Section 7.2.1, which will be briefly discussed here.

Suppose that we have data on L groups or families, the ith family having n_i observations. Let y_{ij} be the observation on the jth member of the ith family. Then, y_{ij} can be represented by a random effects model

$$y_{ij} = \mu + a_i + e_{ij}, \tag{9.1}$$

where μ is the grand mean of all the observations in the population, a_i is the random effect of the ith family, e_{ij} is the error in observing y_{ij}. The random effects a_i are identically distributed with mean 0 and variance σ_a^2, the random errors e_{ij} are identically distributed with mean 0 and variance σ_e^2, and the a_i, e_{ij} are completely independent. The variance of y_{ij} is then given by $\sigma^2 = \sigma_a^2 + \sigma_e^2$, and the covariance between y_{ij} and y_{il} is

$$\text{Cov}\{(\mu + a_i + e_{ij})(\mu + a_i + e_{il})\} = \text{Var}(a_i) = \sigma_a^2.$$

The quantities σ_a^2 and σ_e^2 are called the variance components. These variance components can be estimated from the analysis of variance table (see Section 7.2.1) as follows.

Let $n_0 = (M - \sum_{i=1}^{p} n_i^2 / M)/(p-1)$, where $M = \sum_{i=1}^{p} n_i$ and p is the number of families studied. Then, the following table shows the analysis of variance corresponding to this model.

Source	Degrees of freedom	Sum of squares	Mean square	E(MS)
Between families	$p-1$	SSB	MSB	$n_0 \sigma_a^2 + \sigma_e^2$
Within families	$M-p$	SSW	MSW	σ_e^2

An unbiased estimate of σ_e^2 is $\hat{\sigma}_e^2 = \text{MSW}$ and that of σ_a^2 is $\hat{\sigma}_a^2 = (\text{MSB} - \text{MSW})/n_0$.

The random effects models or the more general mixed effects models are used in longitudinal data analysis. Longitudinal data arise as continuous responses or discrete responses. In what follows we first show methodologies for the analysis of continuous data.

9.3   Linear Mixed Effects models

Longitudinal continuous response data with covariates are usually analyzed on the assumption that the error terms in the regression model are normally distributed.

Consider a longitudinal study involving N subjects, each subject producing a varying number of observations at different time points. Let $Y_i = (Y_{i1}, \ldots, Y_{in_i})^\top$ be the n_i-dimensional vector of observations for the ith subject, $i = 1, \ldots, N$. Assuming an average linear trend for Y as a function of time, a multivariate regression model can be obtained by assuming that the elements Y_{ij} in Y_i satisfy $Y_{ij} = \beta_0 + \beta_1 t_{ij} + \varepsilon_{ij}$, with the assumption that the error components ε_{ij}, $j = 1, \ldots, n_i$, are normally distributed with mean zero. In vector notation, we can write the regression model as $Y_i = X_i \beta + \varepsilon_i$ for an appropriate design matrix X_i, with $\beta^\top = (\beta_0, \beta_1)$ and $\varepsilon_i^\top = (\varepsilon_{i1}, \varepsilon_{i2}, \ldots, \varepsilon_{in_i})$. Using an appropriate variance-covariance matrix V_i for ε_i, we obtain

$$Y_i \sim N(X_i \beta, V_i). \tag{9.2}$$

Note that if the repeated measurements Y_{ij} are assumed to be independent, then $V_i = \sigma^2 I_{n_i}$, where I_{n_i} is the identity matrix of dimension n_i. The above model is based on the assumption that the N subjects are chosen in advance and our purpose is estimate the regression coefficients β_0 and β_1. This is a fixed effects model. The parameters β_0 and β_1 are called population averaged effects.

However, if the N subjects are chosen at random from a population of subjects the fixed effects model fails to account for the differences between the subjects. So, we introduce subject-specific effects, say b_i for the ith subject, into the model. Since the subjects are randomly chosen, b_i is called a random effect having a probability distribution which is typically taken as $N(0, \sigma_b^2)$ (see Section 9.2). The model for Y_{ij} then becomes $Y_{ij} = \beta_0 + \beta_1 t_{ij} + b_i + \varepsilon_{ij}$ and the model for the vector of observations become $Y_i = X_i \beta + b_i + \varepsilon_i$.

For fixed b_i, the distribution of Y_i is

$$Y_i|b_i \sim N(X_i\beta + b_i, V_i). \tag{9.3}$$

If b_i is random and is distributed as $N(0, \sigma_b^2)$, then V_i becomes a matrix with $\sigma^2 + \sigma_b^2$ as the diagonal elements and σ^2 as the off diagonal elements, and the distribution of Y_i comes

$$Y_i \sim N(X_i\beta, V_i). \tag{9.4}$$

Model (9.4) is a mixed effects model. If there is no covariate, this model is identical to the random effects model in Section 9.1.2. However, because of the longitudinal nature of the model, b_i could have covariates which themselves could be correlated. This results in model (4.3) and (4.4) in Molenberghs and Verbeke (2005) and are

$$Y_i|b_i \sim N(X_i\beta + Z_ib_i, \Sigma_i) \tag{9.5}$$

and

$$b_i \sim N(0, D), \tag{9.6}$$

where X_i and Z_i are $n_i \times p$ and $n_i \times q$ known design matrices, respectively, for a p-dimensional vector β of unknown regression coefficients and a q-dimensional vector b_i of subject-specific regression coefficients assumed to be random samples from the q-dimensional normal distribution with mean zero and covariance D, and with Σ_i covariance matrix parameterized through a set of unknown parameters. The components in β are called fixed effects, the components in b_i are called random effects. The fact that the model contains fixed as well as random effects motivates the term mixed effects model (Molenberghs and Verbeke, 2005). Unless the model is fitted in a Bayesian framework (Gelman et al 1995), estimation and inference are based on the marginal distribution for the response vector Y_i. Let $f_i(y_i|b_i)$ and $f(b_i)$ be the density functions corresponding to (9.5) and (9.6), respectively. The marginal density function of Y_i is

$$f_i(y_i) = \int f_i(y_i|b_i)f(b_i)db_i,$$

which can easily be shown to be the density function of an n_i-dimensional normal distribution with mean vector $X_i\beta$ and with covariance matrix $V_i = Z_iDZ_i^\top + \Sigma_i$.

Estimation of the parameters of model (9.4) and (9.5) is based on the marginal model that, for subject i, Y_i is multivariate normal with mean $X_i\beta$ and covariance $V_i(\alpha) = Z_iDZ_i^\top + \Sigma_i$, where α represents the parameters in the covariance matrices D and Σ_i. The classical approach to estimation is based on the method of maximum likelihood (ML). Assuming independence across subjects, the likelihood takes the form

$$L(\theta) = \prod_{i=1}^{N} \left\{ (2\pi)^{-n_i/2}|V_i(\alpha)|^{-\frac{1}{2}} \times \exp\left[-\frac{1}{2}(Y_i - X_i\beta)^\top V_i^{-1}(\alpha)(Y_i - X_i\beta)\right] \right\}. \tag{9.7}$$

Estimation of $\theta^\top = (\beta^\top, \alpha^\top)$ requires joint maximization of (9.7) with respect to all elements in θ. In general, no analytic solutions are available. So, the only direct method of finding the maximum likelihood estimates of the parameters is by using some numerical optimization routines.

Conditionally on α, the maximum likelihood estimator (MLE) of β is given by (Laird and Ware 1982):

$$\widehat{\beta}(\alpha) = \left(\sum_{i=1}^{N} X_i^\top W_i X_i \right)^{-1} \sum_{i=1}^{N} X_i^\top W_i Y_i, \tag{9.8}$$

where W_i equals V_i^{-1}. In practice, α is not known and can be replaced by its MLE $\widehat{\alpha}$. However, one often also uses the so-called restricted maximum likelihood (REML) estimator for α (Thompson, 1962), which allows one to estimate α without having to estimate the fixed effects in β first. It is known from simpler models, such as linear regression models, that, although, the classical ML estimators are biased, the REML estimators avoid the bias (Verbeke and Molenberghs 2000, Section 5.3).

In practice, the fixed effects in β are often of primary interest, as they describe the average evolution in the population. Conditionally on α, the maximum likelihood (ML) estimate for β is given by (9.8), which is normally distributed with mean

$$E\left[\widehat{\beta}(\alpha)\right] = \left(\sum_{i=1}^{N} X_i^\top W_i X_i \right)^{-1} \sum_{i=1}^{N} X_i^\top W_i E[Y_i] = \beta, \tag{9.9}$$

and covariance

$$\mathrm{Var}\left[\widehat{\beta}(\alpha)\right] = \left(\sum_{i=1}^{N} X_i^\top W_i X_i \right)^{-1} \times \left(\sum_{i=1}^{N} X_i^\top W_i \mathrm{Var}(Y_i) W_i X_i \right) \times \left(\sum_{i=1}^{N} X_i^\top W_i X_i \right)^{-1}$$

$$= \left(\sum_{i=1}^{N} X_i^\top W_i X_i \right)^{-1} \tag{9.10}$$

provided that the mean and covariance were correctly specified in our model, i.e., provided that $\mathbf{E}(Y_i) = X_i\beta$ and $\mathrm{Var}(Y_i) = V_i = Z_i D Z_i^\top$.

The random effects model established by Laird and Ware (1982) explicitly model the individual subject effect. Once the distributions of the random effects are specified, inference can be based on maximum likelihood methods. Lindstrom and Bates (1988) also provided details on computational methods for Newton-Raphson and EM algorithms. For the linear random effects model, the generalized least squares based on the marginal distributions of Y_i (after integrating out b_i) is,

$$\hat{\beta} = \left(\sum_{i=1}^{N} X_i^T \Sigma_i^{-1} X_i \right)^{-1} \sum_{i=1}^{N} X_i^T \Sigma_i^{-1} Y_i. \tag{9.11}$$

As for b_i, the Best Linear Unbiased Predictor (BLUP) for b_i is

$$\hat{b}_i = D Z_i^T \Sigma^{-1}(Y_i - X_i\hat{\beta}). \tag{9.12}$$

When the unknown covariance parameters are replaced by their ML or REML estimates, the resulting predictor,

$$\hat{b}_i = \hat{D} Z_i^T \hat{\Sigma}_i^{-1} (Y_i - X_i \hat{\beta}),$$

is often referred to as the "Empirical BLUP" or the "Empirical Bayes" (EB) estimator (see Harville and Jeske, 1992).

Furthermore, it can be shown that,

$$\text{var}(\hat{b}_i) = D Z_i^T \Sigma_i^{-1} Z_i D^T - D Z_i^T \Sigma_i^{-1} X_i (\sum_{i=1}^{N} X_i^T \Sigma_i^{-1} X_i)^{-1} X_i^T \Sigma_i^{-1} Z_i D^T.$$

Let $\hat{H}_i = \hat{\Sigma}_i - Z_i \hat{D} Z_i^\top$. The ith subject's predicted response profile is,

$$\begin{aligned}
\hat{Y}_i &= X_i \hat{\beta} + Z_i \hat{b}_i \\
&= X_i \hat{\beta} + Z_i \hat{D} Z_i^T \hat{\Sigma}_i^{-1} (Y_i - X_i \hat{\beta}) \\
&= (\hat{H}_i \hat{\Sigma}_i^{-1}) X_i \hat{\beta} + (I - \hat{H}_i \hat{\Sigma}_i^{-1}) Y_i.
\end{aligned}$$

That is, the ith subject's predicted response profile is a weighted combination of the population-averaged mean response profile $X_i \hat{\beta}$, and the ith subject's observed response profile Y_i.

Note that $H_i \Sigma_i^{-1}$ measures the relative between-subject variability, while Σ_i is the overall with-subject and between-subject sources of variability. As a result, $\hat{Y}_i$ assigns additional weight to Y_i (the observation from i itself) due to presence of random effects. Note that $X_i \hat{\beta}$ is the estimate at the population level. If $\hat{H}_i$ is large, and the within-subject variability is greater than the between-subject variability, more weight is given to $X_i \hat{\beta}$, the population-averaged mean response profile. When the between-subject variability is greater than the within-subject variability, more weight is given to the ith subject's observed data Y_i. The *lme* function in R package is widely used and numerical solutions can be easily obtained.

9.4 Generalized Linear Mixed Effects Models

We will first focus on random effects model for discrete response data with covariates usually arise for binary response data, data in the form of proportions or count data.

9.4.1 The Logistic Random Effects Models

First we discuss random effects models for binary response data given in Diggle et al. (1994). We consider a random intercept logistic regression model for binary response data with a random effect α_i given by

$$\text{logit} P(Y_{ij} = 1 | \alpha_i) = \beta_0 + \alpha_i + X_{ij}^\top \beta. \tag{9.13}$$

Now, write $\gamma_i = \beta_0 + \alpha_i$ and assume that X_{ij} does not include an intercept term. Then the joint likelihood function for β and the γ_i is proportional to

$$\prod_{i=1}^{N} \exp\left[\gamma_i \sum_{j=i}^{n_i} y_{ij} + \left(\sum_{j=1}^{n_i} y_{ij} X_{ij}^{\top}\right)\beta - \sum_{i=1}^{n_i} \log\{1 + \exp(\gamma_i + X_{ij}^{\top}\beta)\}\right] \tag{9.14}$$

The conditional likelihood for β given the sufficient statistics for the γ_i has the form

$$\prod_{i=1}^{N} \frac{\exp(\sum_{j=1}^{n_i} y_{ij} X_{ij}^{\top}\beta)}{\sum_{R_i} \exp(\sum_{l=1}^{y_{i.}} X_{il}^{\top}\beta)}, \tag{9.15}$$

where $y_{i.} = \sum_{j=1}^{n_i} y_{ij}$ and the index set R_i contains all the $\binom{n_i}{y_{i.}}$ ways of choosing $y_{i.}$ positive responses out of n_i repeated observations. Conditional maximum likelihood estimates of the parameters are to be obtained by maximizing (9.15) with respect to the parameters β.

9.4.2 The Binomial Random Effects Models

Modeling of random effects for binary response data can be extended to sums of independent binary response data as follows: Let $(X_{i1}, \ldots, X_{in_i})$ be a n_i vector of independent Bernoulli (p_i) distributed responses. Then, given p_i, $Y_i = \sum_{j=1}^{n_i} X_{ij}$ has a binomial (n_i, p_i). However, if we assume that p_i is a random variable (effect) having a beta distribution with certain parameters, say, α and β. Then with reparameterization, the unconditional distribution of Y_i is a beta-binomial distribution with mean π and overdispersion parameter ϕ. This was discussed in Chapter 7 (section 7.2.10) extensively.

Exercise 7. Using the results given in Chapter 7 find maximum likelihood estimates of the parameters π and ϕ of the beta-binomial models for the Medium (M) group data given in Table 7.1 and test whether there is significant over-dispersion in the data.

9.4.3 The Poisson Random Effects Models

Count data arise in numerous biological, biomedical and epidemiological investigations. See, for example, Anscombe (1949), Ross and Preece (1985), and Manton et al. (1981). The analysis of such count data is often based upon the assumption of some form of Poisson model. However, when using a Poisson distribution to model such a datasets, it often occurs that the variance of the data exceeds that which would be normally expected. This phenomenon of overdispersion in count data is quite common in practice. See, for example, Bliss and Fisher (1953) and Saha and Paul (2005).

To capture this overdispersion, let $Y \sim \text{Poisson}(\mu)$. However, the Poisson parameter μ may itself vary from experiment to experiment which follows a certain continuous distribution $f(\cdot)$, where, $f(\cdot)$ would be a two-parameter density function. Thus, given μ, $Y \sim \text{Poisson}(\mu)$ and $\mu \sim f$. Then, the unconditional distribution of Y is a two parameter distribution, one of which is the mean parameter m, say and the other is an overdispersion parameter c.

For modeling count data with overdispersion, a popular and convenient model is the negative binomial distribution (see Manton et al., 1981; Breslow, 1984; Engel, 1984; Lawless, 1987; Margolin et al. 1989). For more references on applications of the negative binomial distribution see Clark and Perry (1989). Different authors have used different parameterizations for the negative binomial distribution (see, for example, Paul and Plackett, 1978; Barnwal and Paul, 1988; Piegorsch, 1990; Paul and Banerjee, 1998). Let Y be a negative binomial random variable with mean parameter m and dispersion parameter c. We write $Y \sim NB(m,c)$, which has probability mass function

$$f(y|m,c) = Pr(Y = y|m,c)$$

$$= \frac{\Gamma(y+c^{-1})}{y!\Gamma(c^{-1})} \left(\frac{cm}{1+cm}\right)^y \left(\frac{1}{1+cm}\right)^{c^{-1}}, \tag{9.16}$$

for $y = 0, 1, \cdots$ and $m > 0$. Now, $\text{Var}(Y) = m(1 + mc)$ and $c > -1/m$. Since c can take a positive as well as a negative value, it is called a dispersion parameter, rather than an overdispersion parameter, and with this range of c, $f(y|m,c)$ is a valid probability function. Obviously, when $c = 0$ variance of the $NB(m,c)$ distribution becomes that of the $\text{Poisson}(m)$ distribution. Moreover, it can be shown that the limiting distribution of the $NB(m,c)$ distribution, as $c \to 0$, is the $\text{Poisson}(m)$. The parameter c and its efficient estimation are, therefore, important in practice.

Variety of estimation procedures for the parameters m and c have been developed over the years. Here we give a few of them.

1. The Maximum Likelihood Estimator

Let $Y_1, \ldots, Y_N$ be a random sample from the negative binomial distribution (1.1). Then the log likelihood, apart from a constant, can be written as

$$l = \sum_{i=1}^{N} \left[y_i \ln(m) - \left(y_i + \frac{1}{c}\right)\ln(1+cm) + \sum_{j=0}^{y_i-1} \ln(1+cj) \right]$$

Maximum likelihood estimators of m and c are then obtained by solving the estimating equations

$$\frac{\partial l}{\partial m} = \sum_{i=1}^{N} \left[\frac{y_i}{m} - \frac{1+cy_i}{1+cm} \right] = 0,$$

and

$$\frac{\partial l}{\partial c} = \sum_{i=1}^{N} \left[\frac{1}{c^2}\ln(1+cm) - \frac{y_i-m}{c(1+cm)} - \sum_{j=0}^{y_i-1} \frac{1}{c(1+cj)} \right] = 0,$$

simultaneously (see Piegorsch, 1990). Solution of the first equation provides $\hat{m} = \bar{y}$. Maximum likelihood estimator of the parameter c, denoted by $\hat{c}_{ML}$, is obtained by solving the second equation after replacing m by $\bar{y}$. It can be seen that the restriction

$\hat{c}_{ML} > -1/y_{max}$, where y_{max} is the maximum of the data values $Y_1, \ldots, Y_n$, must be imposed in solving the equation.

2. Method of Moments Estimator

The method of moments estimators of the parameters m and c obtained by equating the first two sample moments with the corresponding population moments are $\hat{m} = \bar{y}$ and

$$\hat{c}_{MM} = \frac{s^2 - \hat{m}}{\hat{m}^2} \tag{9.17}$$

where $\bar{y}$ is the sample mean and $s^2 = \sum_{i=1}^{N} (y_i - \bar{y})^2 / (N-1)$ is the sample variance.

3. The Extended Quasi-Likelihood Estimator

Estimation of the negative binomial parameter k by maximum quasi-likelihood. Biometrics. 45 (1), pp. 309-316. For joint estimation of the mean and dispersion parameters, Nelder and Pregibon (1987) suggested an extended quasi-likelihood, which assumes only the first two moments of the response variable. The extended quasi-likelihood (EQL) and the estimating equations for m and c was given in Clark and Perry (1989). Note our c is the same as α of Clark and Perry (1989). Without giving details we denote the maximum extended quasi-likelihood estimate of c by $\hat{c}_{EQL}$.

4. The Double-Extended Quasi-Likelihood Estimator

Lee and Nelder (2001) introduced double-extended quasi-likelihood (DEQL) for the joint estimation of the mean and the dispersion parameters. The DEQL methodology requires an EQL for Y_i given some random effect u_i and an EQL for u_i from a conjugate distribution given some mean parameter m and dispersion parameter c. The DEQL then is obtained by combining these two EQLs. The random effects u_i's or some transformed variables v_i are then replaced by their maximum likelihood estimates resulting in a profile DEQL. This profile DEQL is the same as the negative binomial log-likelihood with the factorials replaced by the usual Stirling approximations (Lee and Nelder, 2001, Result 5, p. 996). They argued, however, that the Stirling approximation may not be good for small z, so for non-normal random effects, they suggested the modified Stirling approximation

$$\ln \Gamma(z) \simeq \left(z - \frac{1}{2}\right) \ln(z) + \frac{1}{2} \ln(2\pi) - z + \frac{1}{12z}.$$

Omitting details of the derivation, which can be obtained from the authors, the profile DEQL, with the modified Stirling approximation, is

$$p_v^\star(DEQ) = \sum_{i=1}^{N} \left[y_i \ln(m) + \left(y + \frac{1}{c}\right) \ln\left(\frac{1 + cy_i}{1 + cm}\right) - \frac{1}{2} \ln[2\pi(1 + cy_i - \left(y + \frac{1}{2}\right) \ln(y_i) \right.$$
$$\left. - \frac{c^2 y_i}{12(1 + cy_i)} - \frac{1}{12y_i} \right].$$

From this we obtain the maximum profile DEQL estimating equations for m and c as

$$\sum_{i=1}^{N} \left[\frac{y_i}{m(1 + cm)} - \frac{1}{1 + cm} \right] = 0$$

and

$$\sum_{i=1}^{N}\left[\frac{1}{c^2}\ln\left(\frac{1+cm}{1+cy_i}\right)+\frac{y_i-m}{c(1+cm)}-\frac{2cy_i+2-c}{2c^2(1+cy_i)}+\frac{2-c}{2c^2}-\frac{cy_i(2+cy_i)}{12(1+cy_i)^2}\right]=0$$

respectively. The maximum DEQL estimate for m obtained from the first equation above is $\hat{m}=\bar{y}$. The maximum DEQL estimate of c, denoted by $\hat{c}_{DEQL}$, is obtained by iteratively solving the second equation above after replacing m by $\hat{m}=\bar{y}$. Saha and Paul (2005) provided a detailed comparison, by a simulation study, of these estimators.

9.4.4 Examples: Estimation for European Red Mites Data and the Ames Salmonella Assay Data

The European red mites data and the Ames salmonella assay data are now analyzed. The European red mites data do not have any covariate and have 150 observations in the form of frequency distribution (see Table 1, Saha and Paul (2005). The Ames salmonella assay dataset has one covariate and a total of 18 observations. The maximum likelihood estimates of the parameters m and c for the red mites data with the standard errors in parentheses 1.1467(0.1273) and 0.976(0.2629), respectively.

Example 9: Ames Salmonella Assay Data, see Table 2, Saha and Paul, 2005
     The data in Table 2 of Saha and Paul (2005), were originally given by Margoline et al.(1981). The data from an Ames salmonella reverse mutagenicity assay have a response variable Y, the number of revertant colonies observed on each of three replicate plates, and a covariate x, the dose level of quinoline on the plate. We use the regression model given by
     $E(Y_i|x_i)=m_i=\exp[\beta_0+\beta_1 x_i+\beta_2\ln(x_i+10)]$.
     The maximum likelihood estimating equations of the parameters of a general negative binomial regression model are given by Lawless(1987).
     The maximum likelihood estimates of the parameters β_0,β_1,β_2, and c are 2.197628(0.324576), $-0.000980(0.000386)$,   0.312510(0.087892),   and 0.048763(0.028143) respectively.
     It is straightforward to generalize this framework to generalized linear models with random effects. Recall that the GLM relies on the link function h to specify the mean, $\mu=h^{-1}(X^{\top}\beta)$. As an extension to generalized linear models, we can naturally incorporate random effects into the linear predictor. To be brief, we simply replace $X_i\beta$ by $\mathbf{X}_i\beta+Z_i b_i$, in which both X_i and Z_i are covariates. Once the distributions of b_i and η_i (usually multivariate normal) is specified (assumed), likelihood inference can be carried out in principle. Conditional on $(\mathbf{X}_i,Z_i,b_i)$, we have the same framework as the GLM. But b_i are actually "latent" as in linear mixed effects model. In general, approximations are needed in evaluating the marginal likelihood (after integrating the random effects) due to the nonlinear function μ_i in $\theta_i=\mathbf{X}_i\beta+Z_i b_i$. Evaluate the likelihood or how to approximate the likelihood falls into computational statistics. More details can be found in the excellent books (McCulloch and Searle, 2001).

Note that using an additive random effects model leads to a mixture of GLMs. McLachlan (1997) proposed the EM algorithm for the fitting of mixture distributions by maximum likelihood. This essentially opens a very new direction for estimating dispersion parameters for analyzing overdispersed data.

9.5 Transition Models

The marginal approach models the effects on the population rather than on individuals. In many cases, it is more appropriate to model individual changes to fully utilize the information in longitudinal data (see Ware et al., 1988). A few other authors have henceforth extended the marginal GEE approach to a conditional counterpart known as transition models (Zeger and Qaqish, 1988; Liang et al., 1992; Wang, 1996).

Although it is based on the framework of a marginal model, the proposed method relies on the linear functions of past responses as adjusted mean based on the previous responses (residuals).

The transitional model and other conditional quasi-likelihood methods are generally more efficient (Wang 1996). But the efficiency gain hinges on the correct specification of high moments or equivalently, correct specification of the conditional mean and conditional variance structure. However, in many cases, we are only confident about the second-order moment assumption, and it is not possible to specify the fully conditional mean. If the conditional means are indeed linear functions of past responses, the transitional or conditional model has a close connection with the ordinary GEE approach (Wang, 1999c). The proposed model here can, therefore, be treated as a robust version of transitional models.

The parameter estimates of α from a consistent estimation method will converge to α if the correlation is correctly specified. But under misspecification, $\hat{\alpha}$ will converge to some function of α, which will be denoted as γ, and in general, $\gamma \neq \alpha$. The limiting working correlation matrix, $R(\gamma)$, determines the asymptotic efficiency of U_β.

In fact, the working matrix does not even have to converge to a correlation matrix. As long as the limiting matrix is nonsingular, the asymptotic properties of $\hat{\beta}$ still hold. For example, we may use $\alpha = 2$ in AR(1) as a working matrix. But the resulting estimates of β may be very inefficient. This is because the working matrix only plays a role of "weighting" in (4.18). To obtain efficient parameter estimates $\hat{\beta}$, it is desirable to have computationally reliable and statistically robust methods of producing appropriate "working" covariance models.

A very specific class of conditional models are so-called transition models. In a transition model, a measurement y_{ij} in a longitudinal sequence is described as a function of previous outcomes, or history $H_{ij} = (y_{i1}, \ldots, y_{i,j-1})$ (Diggle et al 2002, p.190). One can write a regression model for the outcome y_{ij} in terms of H_{ij}, or alternatively the error term ε_{ij} can be written in terms of previous error terms. In the case of linear models for Gaussian outcomes, one formulation can be translated easily into another one and specific choices give rise to well-known marginal covariance structures such as, for example, AR(1). Specific classes of transition models are also called Markov models (Feller 1968). The order of a transition model is the number of previous measurements that is still considered to influence the current one. A model is called stationary if the functional form of the dependence is the same regardless of the actual time at which it occurs. An example of a stationary first-order autoregressive model for continuous data is:

$$y_{i1} = X_{i1}^{\top}\beta + \varepsilon_{i1}, \tag{9.18}$$

$$y_{ij} = X_{ij}\beta + \alpha y_{ij-1} + \varepsilon_{ij}. \tag{9.19}$$

Assuming $\varepsilon_{i1} \sim N(0,\sigma^2)$ and $\varepsilon_{ij} \sim N(0,\sigma^2(1-\alpha^2))$, after some simple algebra, we can obtain: $\text{var}(Y_{ij}) = \sigma^2$ and $\text{cov}(Y_{ij}, Y_{ij'}) = \alpha^{|j'-j|}\sigma^2$. In other words, this model produces a marginal multivariate normal model with AR(1) variance-covariance matrix. It makes most sense for equally spaced outcomes, of course. Upon including random effects into (9.1), and varying the assumptions about the autoregressive structure, it is clear that the general linear mixed-effects model formulation with serial correlation encompasses wide classes of transition models. There is a relatively large literature on the direct formulation of Markov models for binary, categorical, and Poisson data as well (Diggle et al 2002). For outcomes of a general type, generalized linear model ideas can be followed to formulate transition models.

We extend the generalized linear models (GLMs) for describing the conditional distribution of each response y_{ij} as a function of past response $y_{ij-1},\ldots,y_{i1}$ and covariates X_{ij}. For example, the probability of a child having a obesity problem at time t_{ij} depends not only on explanatory variables, but also on the obesity status at time t_{ij-1}. We will focus on the case where the observation times t_{ij} are equally spaced. To simplify notation, we denote the history for subject i at visit j by $H_{ij} = (y_{i1}, y_{i2}, \ldots, y_{ij-1})$. As above, we will continue on the past and present values of the covariates without explicitly listing them.

The most useful transition models are Markov chains for which the conditional distribution of y_{ij} given H_{ij} depends only on the q prior observations $y_{ij-1},\ldots,y_{ij-q}$, The integer q is referred to as the model order.

A transition model specifies a GLM for the conditional distribution of y_{ij} given the past response, H_{ij}. The form of the conditional GLM is

$$f(y_{ij}|H_{ij}) = \exp\left\{[y_{ij}\theta_{ij} - \psi(\theta_{ij})]/\phi + c(y_{ij},\phi)\right\} \tag{9.20}$$

for the known functions $\psi(\theta_{ij})$ and $c(y_{ij},\phi)$. The conditional mean and variance are

$$\mu_{ij}^c = \mathbf{E}(Y_{ij}|H_{ij}) = \psi'(\theta_{ij}) \text{ and } v_{ij}^c = Var(Y_{ij}|H_{ij}) = \psi''(\theta_{ij})\phi,$$

where ϕ is an overdispersion parameter. The only difference is that, by including H_{ij}, an outcome is described in terms of its predecessors. A function of the mean components is equated to a linear function of the predictors:

$$\eta_{ij}(\mu_{ij}^c) = X_{i1}^{\top}\beta + \kappa(H_{ij},\beta,\alpha), \tag{9.21}$$

where $\kappa(\cdot)$ is a function, often linear, of the history.

In words, the transition model expresses the conditional mean μ^c_{ij} as a function of both the covariates X_{ij} and of the past responses $y_{i1}, y_{i2}, \ldots, y_{ij-q}$. Past responses or functions thereof are simply treated as additional explanatory variables. The following examples with different link functions illustrate the range of transition models which are available.

Linear link—a linear regression with autoregressive errors for Gaussian data (Tsay, 1984) is a Markov model. It has the form

$$Y_{ij} = x_{ij}^\top \beta + \sum_{r=1}^{q} \alpha_r (Y_{ij-r} - x_{ij-r}^\top \beta) + \varepsilon_{ij}, \tag{9.22}$$

where the ε_{ij} are independent, mean-zero, Gaussian innovations. Note that the present observation, Y_{ij}, is a linear function of x_{ij} and of the earlier deviation $Y_{ij-r} - x_{ij-r}^\top \beta$, $r = 1, \ldots, q$.

Logit link—an example of a logistic regression model for binary responses that comprises a first order Markov chain (Cox and Snell, 1989; Korn and Whittemore, 1979, see) which assumes that y_{ij} ($j > 1$) is independent of earlier observations given the previous observation y_{ij-1} is

$$\text{logit} P(Y_{ij} = 1 | H_{ij}) = X_{ij} \beta + Y_{ij-1} \alpha. \tag{9.23}$$

This model is of the stationary first-order autoregressive type. Evaluating (9.23) to $y_{i,j-1} = 0$ and $y_{i,j-1} = 1$, respectively, produces the so-called transition probabilities between occasions $j - 1$ and j. In this model, when there would be no covariates, these would be constant across the population. When there are time-independent covariates only, the transition probabilities change in a relatively straightforward way with level of covariate. For example, a different transition structure may apply to the standard and experimental arms in a two-armed clinical study. A simple extension to a model of order q has the form

$$\text{logit} P(Y_{ij} = 1 | H_{ij}) = X_{ij} \beta_q + \sum_{r=1}^{q} Y_{ij-r} \alpha_r. \tag{9.24}$$

The notation β_q indicates that the value and interpretation of the regression coefficients changes with the Markov order, q.

Log-link— with count data we assume a log-linear model where Y_{ij} given H_{ij} follows a Poisson distribution. Zeger and Qaqish (1988) discussed a first order Markov chain with $f = \alpha \left\{ \log(y^*_{ij-1}) - X_{ij-1}^\top \beta \right\}$, where $y^*_{ij} = \max(y_{ij}, c)$, $0 < c < 1$. This leads to

$$\mu^c_{ij} = \mathbf{E}(Y_{ij} | H_{ij}) = \exp(X_{ij}^\top \beta) \left(\frac{y^*_{ij-1}}{\exp(X_{ij-1}^\top \beta)} \right)^\alpha. \tag{9.25}$$

The constant c prevents $y_{ij} = 0$ from being an absorbing state whereas $y_{ij} = 0$ forces all future response to be 0. Note when $\alpha > 0$, we have an increased expectation, μ^c_{ij}, when the previous outcome, y_{ij-1}, exceeds $\exp(X_{ij-1}^\top \beta)$. When $\alpha < 0$, a higher value at t_{ij-1} causes a lower value at t_{ij}.

In the linear regression model, the transition model can be formulated with $f_r = \alpha_r(y_{ij-r} - X_{ij-1}^\top \beta)$ so that $\mathbf{E}(Y_{ij}) = X_{ij}^\top \beta$ for different values of q. In the logistic and log-linear cases, it is difficult to formulate models in such a way that β has the same meaning for different assumptions about the time dependence. When β is the scientific focus, the careful data analyst should exam the sensitivity of the substantive findings to the choice of time dependence model.

9.6 Fitting Transition Models

The contribution to the likelihood for the ith subject can be written as:

$$L_i(y_{i1}, \cdots y_{in_i}) = f(y_{i1}) \prod_{j=2}^{n_i} f(y_{ij}|H_{ij})$$

$$= f(y_{i1}, \cdots, y_{iq}) \prod_{j=q+1}^{n_i} f(y_{ij}|y_{ij-1}, \cdots, y_{ij-q}). \qquad (9.26)$$

The latter decomposition is relevant as the history H_{ij} contains the Q immediately preceding measurements. It is now clear that the product in (9.13) yields n_i independent univariate GLM contributions. Clearly, a separate model may need to be considered for the first q measurements, as these are left undescribed by the conditional GLM. In the linear model we assume that y_{ij} given H_{ij} follows a normal distribution. If $y_{i1}, \ldots, y_{iq}$ are also multivariate normal and the covariance structure for the y_{ij} is weakly stationary, the marginal distribution $f(y_{ij}, \cdots, y_{iq})$ can be fully determined from the conditional distribution model without additional unknown parameters. Hence, full maximum likelihood estimation can be used to fit normal autoregressive models. See Tsay (1984) and references therein for details.

In the logistic and log-linear cases, $f(y_{i1}, \cdots, y_{iq})$ is not determined from the GLM assumption about the conditional model, and the full likelihood is unavailable. An alternative is to estimate β and α by maximizing the conditional likelihood

$$\prod_{i=1}^{N} f(y_{iq+1}, \cdots, y_{in_i}|y_{i1}, \cdots, y_{iq}) = \prod_{i=1}^{N} \prod_{j=q+1}^{n_i} f(y_{ij}|H_{ij}). \qquad (9.27)$$

When maximizing (9.27) there are two distinct cases to consider. In the first, $f_r(H_{ij}; \alpha, \beta) = \alpha_r f_r(H_{ij})$ so that $h(u_{ij}^c) = X_{ij}^\top \beta + \sum_{r=1}^{s} \alpha_r f_r(H_{ij})$. Here, $h(u_{ij}^c)$ is a linear function of both β and $\alpha = (\alpha_1, \ldots, \alpha_s)$ so that the estimation proceeds as in GLMs for independent data. We simply regress y_{ij} on the $(p+s)$ dimensional vector of extended explanatory variables $(x_{ij}, f_1(H_{ij}), \ldots, f_s(H_{ij}))$.

The second case occurs when the functions of past responses include both α and β. Examples are the linear and log-linear models discussed above. To derive an estimation algorithm for this case, note that the derivative of the log conditional likelihood or conditional score function has the form

$$U^c(\delta) = \sum_{i=1}^{N} \sum_{j=q+1}^{n_i} \frac{\partial u_{ij}^c}{\partial \delta} v_{ij}^c (y_{ij} - u_{ij}^c) = 0, \qquad (9.28)$$

where $\delta = (\beta, \alpha)$. This equation is the conditional analog of the GLM score equation. The derivative $\partial u_{ij}^c / \partial \delta$ is analogous to X_{ij} but it can depend on α and β. We can still formulate the estimation procedure as an iterative weighted least squares as follows. Let Y_{ij} be the $(n_i - q)$ vector of responses for $j = q+1, \ldots, n_i$ and μ_{ij}^c its expectation given H_{ij}. Let X_i^* be an $(n_i - q) \times (p + s)$ matrix with kth row $\partial u_{iq+k} / \partial \delta$ and $W_{ij} = \text{diag}(1/v_{iq+1}^c, \cdots, 1/v_{in_i}^c)$ and $(n_i - q) \times (n_i - q)$ diagonal weighting matrix. Finally, let $Z_i = X_i^* \hat{\delta} + (Y_i - \hat{\mu}^c)$. Then, an updated $\hat{\delta}$ can be obtained by iteratively regressing Z on X^* using weights W.

When the correct model is assumed for the conditional mean and variance, the solution $\hat{\delta}$ asymptotically, as N goes to infinity, follows a Gaussian distribution with mean equal to the true value, δ, and $(p + s) \times (p + s)$ variance matrix

$$V_{\hat{\delta}} = \left(\sum_{i=1}^{N} X_i^* W_i X_i^{*\top} \right)^{-1}. \tag{9.29}$$

The variance $V_{\hat{\delta}}$ depends on β and α. A consistent estimate, $\hat{V}_{\hat{\delta}}$ is obtained by replacing β and α by their estimates $\hat{\beta}$ and $\hat{\alpha}$. Hence, a 95 percent confidence interval for β_1 is $\hat{\beta}_1 \pm 2 \sqrt{\hat{V}_{\hat{\delta}}}$, where $\hat{V}_{\hat{\delta}}$ is the element in the first row and column of $\hat{V}_{\hat{\delta}}$.

If the conditional mean is correctly specified but the conditional variance is not, we still obtain consistent inferences about δ by using the robust variance from equation (A.6.1) in Diggle et al. (1994, pp 245). Here it takes the form

$$V_R = \left(\sum_{i=1}^{N} X_i^{*\top} W_i X_i^* \right)^{-1} \left(\sum_{i=1}^{N} X_i^{*\top} W_i V_{Ti} W_i X_i^* \right) \left(\sum_{i=1}^{N} X_i^{*\top} W_i X_i^* \right)^{-1}. \tag{9.30}$$

A consistent estimate $\hat{V}_R$ is obtained by replacing $V_{Ti} = \text{Var}(Y_i | H_i)$ in the equation above by its estimates $(Y_i - \hat{\mu}_i^c)(Y_i - \hat{\mu}_i^c)^\top$.

Interestingly, use of the robust variance will often give consistent confidence intervals for $\hat{\delta}$ even when the Markov assumption is violated. However, in that situation, the interpretation of $\hat{\delta}$ is questionable since $\mu_{ij}^c(\hat{\delta})$ is not the conditional mean of Y_{ij} given H_{ij}.

9.7   Transition Model for Categorical Data

This section discusses Markov chain regression models for categorical responses observed at equally spaced intervals. We begin with logistic models for binary responses and then briefly consider extensions to multinominal and ordered categorical outcomes.

A first order binary Markov chain is characterized by the transition matrix

$$\begin{pmatrix} \pi_{00} & \pi_{01} \\ \pi_{10} & \pi_{11} \end{pmatrix},$$

where $\pi_{ab} = p(Y_{ij} = b | Y_{ij-1} = a)$, $a, b = 0, 1$. For example, π_{01} is the probability that

$Y_{ij} = 1$ when the previous response is $Y_{ij-1} = 0$. Note that each row of a transition matrix sums to one. As its name implies, the transition matrix records the probabilities of making each of the possible transitions from one visit to the next.

In the regression setting, we model the transition probabilities as functions of covariates X_{ij}. A very general model uses a separate logistic regression for $P(Y_{ij=1}|Y_{ij-1} = y_{ij})$, $y_{ij} = 0, 1$. That is, we assume that

$$\text{logit} P(Y_{ij} = 1|Y_{ij-1} = 0) = X_{ij}^\top \beta_0$$

and

$$\text{logit} P(Y_{ij} = 1|Y_{ij-1} = 1) = X_{ij}^\top \beta_1,$$

where β_0 and β_1 may differ. In words, this model assumes that the effects of explanatory variables will differ depending on the previous response. A more concise form for the same model is

$$\text{logit} P(Y_{ij} = 1|Y_{ij-1} = y_{ij-1}) = X_{ij}^\top \beta_0 + y_{ij-1} X_{ij}^\top \alpha, \qquad (9.31)$$

so that $\beta_1 = \beta_0 + \alpha$. An advantage of this form is that we can easily test whether simpler models fit the data equally well. For example, we can test whether $\alpha = (\alpha_0, 0)$ indicates that the covariates have the same effect on the response probability whether $y_{ij-1} = 0$ or $y_{ij-1} = 1$. Alternately, we can test whether a more limited subset of α is zero indication that the associated covariates can be dropped from the model. Each of these alternatives is nested within the saturated model so that standard statistical method for nested models can be applied.

In many problems, a higher order Markov chain may be needed. The second order model has transition matrix

		Y_{ij}	
Y_{ij-2}	Y_{ij-1}	0	1
0	0	π_{000}	π_{001}
0	1	π_{010}	π_{011}
1	1	π_{110}	π_{111}

Here, $\pi_{abc} = P(Y_{ij} = c|Y_{ij-2} = a, Y_{ij-1} = b)$; for example π_{011} is the probability that $Y_{ij} = 1$ given $Y_{ij-2} = 0$ and $Y_{ij-1} = 1$. We could fit four separate logistic regression models, one for each of the four possible histories (Y_{ij-2}, Y_{ij-1}) namely $(0,0), (0,1), (1,0)$, and $(1,1)$ with regression coefficients $\beta_{00}, \beta_{01}, \beta_{10}, \beta_{11}$ respectively. But it is again more convenient to write a single equation as follows

$$\text{logit} P(Y_{ij} = 1|Y_{ij-2} = y_{ij-2}, Y_{ij-1} = y_{ij-1})$$
$$= X_{ij}^\top \beta + y_{ij-1} X_{ij} \alpha_1 + y_{ij-2} X_{ij}^\top \alpha_2 + y_{ij-1} y_{ij-2} X_{ij}^\top \alpha_3. \qquad (9.32)$$

By plugging in the different values for y_{ij-2} and y_{ij-1}, we obtain $\beta_{00} = \beta, \beta_{01} = \beta + \alpha_1$, $\beta_{10} = \beta + \alpha_2$, and $\beta_{11} = \beta + \alpha_1 + \alpha_2 + \alpha_3$. We would again hope that a more parsimonious model fits the data equally well so many of the components of the α_i would be zero.

An important special case occurs when there are no interactions between the past responses, y_{ij-1} and y_{ij-2} and the explanatory variables, that is, when all elements of the α_i are zero except the intercept term. In this case, the previous responses affect the probability of a positive outcome but the effects of the explanatory variables are the same regardless of the history. Even in this situation, we must still choose between Markov models of different order. For example, we might start with a third order model which can be written in the form

$$\text{logit} P(Y_{ij} = 1 | Y_{ij-3} = y_{ij-3}, Y_{ij-2} = y_{ij-2}, Y_{ij-1} = y_{ij-1})$$
$$= x'_{ij}\beta + y_{ij-1}\alpha_1 + y_{ij-2}\alpha_2 + y_{ij-3}\alpha_3 + y_{ij-1}y_{ij-2}\alpha_4 + y_{ij-1}y_{ij-3}\alpha_5$$
$$+ y_{ij-2}y_{ij-3}\alpha_6 + y_{ij-1}y_{ij-2}y_{ij-3}\alpha_7. \tag{9.33}$$

A second order model can be used if the data are consistent with $\alpha_3 = \alpha_5 = \alpha_6 = \alpha_7 = 0$; a first order model is implied if $\alpha_i = 0$ for $j = 2, \ldots, 7$. As with any regression coefficients, the interpretation and value of β depends on the other explanatory variables in the model, in particular on which previous responses are included. When inference about β are the scientific focus, it is essential to check their sensitivity to the assumed order of Markov regression model. When the Markov transition model is correctly specified, the transition events are uncorrelated so that ordinary logistic regression can be used to estimate regression coefficients and their standard error. However, there may be circumstances when we choose to model $P(Y_{ij} | Y_{ij-1}, \cdots, Y_{ij-q})$ even though it does not equal $P(Y_{ij} | H_{ij})$. For example, suppose there is heterogeneity across people in the transition matrix due to unobserved factors, so that a reasonable model is

$$\text{logit} P(Y_{ij} = 1 | Y_{ij-1}, U_i = y_{ij-1}) = (\beta_0 + U_i) + X_{ij}^{\top}\beta + \alpha y_{ij-1}, \tag{9.34}$$

where $U_i \sim N(0, \sigma^2)$. We may still wish to estimate the population-averaged transition matrix, $P(Y_{ij} | Y_{ij-1} = y_{ij-1})$. But here the random intercept U_i makes the transitions for a person correlated. Correct inference about the population-average coefficients can be drawn using the GEE approach.

To capture the baseline heterogeneity across subjects (Albert and Follmann, 2003) for the first order binary Markov chain, a separate random intercept effect logistic regression model can be used as

$$\text{logit}(P_{01}(u_i)) = \text{logit}[P(Y_{ij} = 1 | Y_{ij-1} = 0, u_i)] = X_{ij}^{\top}\beta_{01} + u_i$$
$$\text{logit}(P_{10}(u_i)) = \text{logit}[P(Y_{ij} = 0 | Y_{ij-1} = 1, u_i)] = X_{ij}^{\top}\beta_{10} + vu_i$$

from which we obtain

$$P_{01}(u_i) = \frac{\exp(X_{ij}^{\top}\beta_{01} + u_i)}{1 + \exp(X_{ij}^{\top}\beta_{01} + u_i)}$$

$$P_{10}(u_i) = \frac{\exp(X_{ij1}^{\top}\beta_{10} + vu_i)}{1 + \exp(X_{ij}^{\top}\beta_{10} + vu_i)}$$

where β_{01} and β_{10} are regression parameters. The parameter v represents the association between $P_{01}(u_i)$ and $P_{10}(u_i)$. Note that $P_{01} + P_{00} = 1 = P_{11} + P_{10} = 1$. To see the effect of v for some fixed parameters see Figure 3 (Albert and Follmann, 2003, pp 104). Thus, the model for y_{ij} given x_{ij}, $y_{i,j-1}$, u_i, and $\beta = (\beta_0, \beta_1)$ can be written as

$$
f(y_{ij}|X_{ij}, z_{ij}, y_{i,j-1}, u_i, \theta) = \begin{cases} P_{01}^{y_{ij}}(u_i)(1 - P_{01}(u_i))^{1-y_{ij}} & \text{if } y_{i(j-1)} = 0 \\ P_{10}^{1-y_{ij}}(u_i)(1 - P_{10}(u_i))^{y_{ij}} & \text{if } y_{i(j-1)} = 1. \end{cases} \tag{9.35}
$$

9.8 Further reading

Markov transition models have been studied by several authors. For example, Korn and Whittemore (1979) modeled the probability of occupying the current state using the previous state. Wu and Ware (1979) assumed one binary event (e.g., death) though the covariate information as time passes before the event. Regier (1968) reparametrized the two-state transition matrix to include a parameter which is the ratio of the odds of staying in state 0 to the odds of staying in state 1. Zeger and Qaqish (1988) discussed a class of Markov regression models for time-series by using a quasi-likelihood approach. Lee and Kim (1998) analysed correlated panel data using a continuous-time Markov Model. Zeng and Cook (2007) proposed a estimation method based on joint transitional models for multivariate longitudinal binary data using GEE2. Deltour, Richardson, and Le Hesran (1999) used stochastic algorithms for Markov models estimation with intermittent missing data. Albert (2000) developed a transitional model for longitudinal binary data subject to nonignorable missing data and proposed an EM algorithm for parameter estimation. In Albert and Follmann (2003), an extended version of the Markov transition model was proposed to handle nonignorable missing values in a binary longitudinal data set.

Chapter 10

Handing High Dimensional Longitudinal Data

10.1 Introduction

Many variables are often collected in high-dimensional longitudinal data. The inclusion of redundant variables may reduce accuracy and efficiency for both parameter estimation and statistical inference. The traditional criteria introduced in Chapter 5 are all subset methods, and these methods become computationally intensive when dimension of the variables is moderately large. Therefore, it is important to find new methodology for variable selection in analysis of high-dimensional longitudinal data. The penalized loss function (or $-2\log$ likelihood function) methods have been widely used to select variables in regression models for the independent data $\{(X_i, Y_i) : i = 1, \cdots, N\}$. The penalized loss function method is composed of a loss function and a penalty function, that is $Q(\beta) = \sum_{i=1}^{N} l_i(\beta) + N \sum_{j=1}^{p} P_\lambda(|\beta_j|)$, where $P_\lambda(|\beta|)$ is a penalty function, and λ is a tuning parameter which controls the sparseness of the regression parameters β, the commonly used penalty functions $P_\lambda(|\theta|)$ include,

- Hard penalty function: $P_\lambda(|\theta|) = \lambda^2 - (|\theta| - \lambda)^2 I(|\theta| < \lambda)$.

- Least absolute shrinkage and selection operator (LASSO) (Tibshirani, 1996): $P_\lambda(|\theta|) = \lambda|\theta|$.

- Smoothly clipped absolute deviation penalty (SCAD) (Fan and Li, 2001):

$$P'_\lambda(|\theta|) = \lambda \left\{ I(|\theta| \le \lambda) + \frac{(a\lambda - |\theta|)_+}{(a-1)\lambda} I(|\theta| > \lambda) \right\},$$

  where $a > 2$ and is proposed taking a value of 3.7 by Fan and Li (2001).

- Adaptive penalty function (ALASSO) (Zou, 2006): $P_\lambda(|\theta|) = \lambda|\theta|/|\hat\theta|^\gamma$, where $\hat\theta$ is a consistent estimator of θ and $\gamma > 0$.

- Elastic net (EN) penalty (Zou and Hastie, 2005) introduced as a mixing penalty to effectively select grouped variables: $P_\lambda(|\theta|) = \lambda_1\|\theta\|_1 + \lambda_2\|\theta\|_2^2$.

The penalized methods can compress the coefficients of unrelated variables toward zero and obtain the parameter estimates at the same time. Fan and Li (2001) proposed that a good penalty function should result in an estimator with three properties: unbiasedness, sparsity, and continuity. Their studies found that the excessive

DOI: 10.1201/9781315153636-10

compression of the coefficients in LASSO would cause large deviations and then proposed a smoothly clipped absolute deviation (SCAD) penalty function to solve this problem. SCAD is more stable than LASSO and can reduce the computational complexity. Furthermore, the adaptive LASSO permits different weights for different coefficients and has oracle properties.

In this chapter, we will introduce several advanced methods for simultaneous parameter estimation and variable selection in marginal models with longitudinal data. Assume that $\mu_{ij} = g(X_{ij}^T\beta)$, where $g(\cdot)$ is a specified function and β is an unknown parameter vector, and we assume that β is sparse and partition β as $(\beta_1^T, \beta_2^T)^T$ with $\beta_1 \in \mathbb{R}^d$ and $\beta_2 \in \mathbb{R}^{p-d}$. Suppose that the true values of parameters β are $\beta_T = (\beta_1^{*T}, \mathbf{0}_2^T)^T$. We aim to identify the variables with zero coefficients β_2 consistently, and simultaneously estimate the nonzero coefficients β_1.

10.2 Penalized Methods

In this section, we will introduce several penalized methods for variable selection in marginal models.

10.2.1 Penalized GEE

The GEE method introduced in Chapter 4 is the most popular method in estimating parameters in marginal models. In this section, we will introduce a penalized GEE proposed by Wang et al. (2012a) for variable selection. Let

$$U_N(\beta) = \sum_{i=1}^{N} D_i^T A_i^{-1/2} R_i^{-1}(\alpha) A_i^{-1/2} (Y_i - \mu_i).$$

The penalized GEE is

$$Q_N(\beta) = U_N(\beta) - NP'_\lambda(|\beta|)\text{sign}(\beta), \qquad (10.1)$$

where $P'_\lambda(|\beta|) = (P'_\lambda(|\beta_1|), \cdots, P'_\lambda(|\beta_p|))^T$ is a p dimensional derivative function of the penalty function $P_\lambda(|\beta|)$ encouraging sparsity in β, and $\text{sign}(\beta) = (\text{sign}(\beta_1), \cdots, \text{sign}(\beta_p))^T$ is an indicator vector with $\text{sign}(t) = 1$ for $t > 0$, $\text{sign}(t) = -1$ for $t < 0$ and $\text{sign}(t) = 0$ for $t = 0$. In Equation (10.1), the tuning parameter $\lambda > 0$ controls the complexity of the model.

Wang et al. (2012a) proposed combining the Minimization-Maximization (MM) algorithm by Hunter and Li (2005) with the Fisher-scoring algorithm to solve the penalized GEE (10.1). For a small $\epsilon > 0$, the MM algorithm suggests that the penalized estimator $\hat{\beta}$ approximately satisfies the following equation:

$$U_{Nj}(\hat{\beta}) - NP'_\lambda(|\hat{\beta}_j|)\text{sign}(\hat{\beta}_j)\frac{|\hat{\beta}_j|}{|\hat{\beta}_j| + \epsilon} = 0, \qquad (10.2)$$

where $U_{Nj}(\hat{\beta})$ is the jth element of $U_N(\hat{\beta})$, and $\hat{\beta}_j$ is an estimator of β_j for $i = 1, \cdots, p$. Applying the Fisher-scoring algorithm to (10.2), we can obtain the following iterative formula for the penalized GEE:

$$\hat{\beta}^{(k+1)} = \hat{\beta}^{(k)} + [H(\hat{\beta}^{(k)}) + NE(\hat{\beta}^{(k)})]^{-1}[U_N(\hat{\beta}^{(k)}) - NE(\hat{\beta}^{(k)}) \times \hat{\beta}^{(k)}], \qquad (10.3)$$

where

$$H(\hat{\beta}^{(k)}) = \sum_{i=1}^{N} X_i^T A_i^{1/2}(\hat{\beta}^{(k)}) R_i^{-1}(\alpha) A_i^{1/2}(\hat{\beta}^{(k)}) X_i,$$

$$\mathbf{E}(\hat{\beta}^{(k)}) = \text{diag}\left(\frac{P'_\lambda(|\hat{\beta}_1^{(k)}|+)}{|\hat{\beta}_1^{(k)}| + \epsilon}, \cdots, \frac{P'_\lambda(|\hat{\beta}_p^{(k)}|+)}{|\hat{\beta}_p^{(k)}| + \epsilon}\right).$$

Given a selected tuning parameter λ and an initial value of β, such as the estimate obtained from the GEE with an independence working matrix under a full model, update the estimate of β via equation (10.3) until the algorithm converges.

The tuning parameter λ controls the spareness of the regression parameters, and it is important to select an appropriate λ. The cross-validation method is very popular for selecting the tuning parameter. The k-fold cross-validation procedure is as follows: Denote the full dataset by $\mathscr{T}$ and randomly split the data into k nonoverlapping subsets of approximately equal size. Denote cross-validation training and test set by $\mathscr{T} - \mathscr{T}^v$ and $\mathscr{T}^v$, respectively. Obtain the estimator $\hat{\beta}^v$ of β using the training set $\mathscr{T} - \mathscr{T}^v$. Form the cross validation criterion as

$$CV(\lambda) = \sum_{v=1}^{k} \sum_{(y_{ij}, x_{ij}) \in \mathscr{T}^v} l(y_{ij}, x_{ij}, \hat{\beta}^{(v)}),$$

where k usually takes a value of 5 or 10, and $l(\cdot)$ is a loss function. Wang et al. (2012) proposed taking the negative log likelihood of exponential family distribution under a working independence assumption as the loss function. The best tuning parameter is selected by minimizing $CV(\lambda)$ over a fine grid of λ. From the MM algorithm, a sandwich covariance formula can be used to estimate the asymptotic covariance of $\hat{\beta}$:

$$\widehat{\text{Cov}}(\hat{\beta}) = [H(\hat{\beta}) + NE(\hat{\beta})]^{-1} M(\hat{\beta})[H(\hat{\beta}) + NE(\hat{\beta})]^{-1},$$

where

$$M(\hat{\beta}) = \sum_{i=1}^{N} X_i^T A_i^{1/2}(\hat{\beta}) \hat{R}_i^{-1}(\alpha) \epsilon_i^T(\hat{\beta}) \epsilon_i(\hat{\beta}) \hat{R}_i^{-1} A_i^{1/2}(\hat{\beta}) X_i,$$

where $\epsilon_i(\hat{\beta}) = A_i^{-1/2}(\hat{\beta})(Y_i - \mu_i(\hat{\beta}))$. The tuning parameter can also be selected by minimizing the determinant or trace of the covariance matrix $\widehat{\text{Cov}}(\hat{\beta})$. The correlation parameters α in the working correlation matrix can be estimated using the moment method introduced in Chapter 4.

The penalty function $P_\lambda(|\beta|)$ can be SCAD or ALASSO. The resulting estimators from (10.1) are unbiased, consistent, and have oracle properties (Wang et al., 2012). A *PGEE* package in statistical software R can be used to fit the penalized generalized estimating equations by SCAD penalty to longitudinal data with high-dimensional covariates.

10.2.2 *Penalized Robust GEE-type Methods*

The penalized GEE is sensitive to outliers and, therefore, some researchers proposed penalized robust GEE-type methods. For example, Fan et al. (2012) and Lv et al. (2015) proposed a robust variable selection approach based on robust generalized estimating functions that incorporate the correlation structure for longitudinal data. Consider the following robust estimating equations:

$$U_R(\beta) = \sum_{i=1}^{N} D_i^{\mathrm{T}} V_i^{-1} h_i(\mu_i(\beta)) = 0, \tag{10.4}$$

where $V_i = R_i(\alpha) A_i^{1/2}$, $h_i(\mu_i(\beta)) = W_i(\psi(A_i^{-\frac{1}{2}}(Y_i - \mu_i)) - C_i)$ in which $C_i = \mathrm{E}(\psi(A_i^{-\frac{1}{2}}(Y_i - \mu_i)))$ is used to ensure Fisher consistency of the estimator. For the Gaussian distribution and symmetric Huber function, the correction term C_i is exactly equal to zero.

The function $\psi(\cdot)$ is selected to downweight the influence of outliers in the response variable. Fan et al. (2012) chose the Huber function $\psi_c(x) = \min\{c, \max(-c, x)\}$, where the tuning constant c is chosen to give a certain level of asymptotic efficiency at the underlying distribution. Lv et al. (2015) proposed using a bounded exponential score function $\psi(t) = \exp(-t^2/\gamma)$, where γ downweights the influence of an outlier on the estimators. The weighted function $W_i = \mathrm{diag}(w_{i1}, w_{i2}, \cdots, w_{in_i})$ is used to downweight the effect of leverage points and can be chosen using the Mahalanobis distance function given as

$$w_{ij} = w_{ij}(X_{ij}) = \min\left\{1, \left\{\frac{b_0}{(X_{ij} - m_x)^{\mathrm{T}} S_x^{-1}(X_{ij} - m_x)}\right\}^{\frac{r}{2}}\right\},$$

where $r \geq 1$ and b_0 is the 0.95 quantile of a χ^2 distribution with p degrees of freedom, m_x and S_x are some robust estimates of location and scatter of X_{ij}, such as the median of X_{ij} or median absolute deviance of X_{ij}. If we set $\psi(x) = x$ and $w_{ij} = 1$, then $U_R(\beta)$ in (10.4) will be equal to $U_N(\beta)$.

To select important covariate variables and estimate them simultaneously, the penalized method introduced in §10.2.1 can be used to select important variables by adding a penalty function in $U_R(\beta)$. The penalized robust estimating equations are given as follows:

$$U_{PR}(\beta) = U_R(\beta) - N P_\lambda'(|\beta|)\mathrm{sign}(\beta) = 0. \tag{10.5}$$

The MM iterative algorithm can be used to solve Equation (10.5). The correlation parameters and the scale parameter ϕ need to be estimated before solving (10.5). Let $\hat{e}_{ij} = (y_{ij} - \mu_{ij}(\hat{\beta}))/\sqrt{v(\mu_{ij}(\hat{\beta}))}$ be Pearson residuals. We can obtain a robust estimate of ϕ through the median absolute deviation

$$\hat{\phi} = \{1.483 \mathrm{median}|\hat{e}_{ij} - \mathrm{median}\{\hat{e}_{ij}\}|\}^2. \tag{10.6}$$

For a given score function $\psi(\cdot)$, we obtain the estimates of α by the method of Wang et al. (2005). When $R(\alpha)$ is an exchangeable correlation matrix, α can be estimated by

$$\hat{\alpha} = \frac{1}{NH} \sum_{i=1}^{N} \frac{1}{n_i(n_i-1)} \sum_{j \neq k} \psi(e_{ij})\psi(e_{ik}), \qquad (10.7)$$

where the summation is over all pairs $(j \neq k)$ and H is the mean of $\psi^2(e_{ij})$, $j = 1, \cdots, n_i$, $i = 1, \cdots, N$. When $R(\alpha)$ is an AR(1) correlation matrix, α can be estimated by

$$\hat{\alpha} = \frac{1}{NH} \sum_{i=1}^{N} \frac{1}{n_i-1} \sum_{j \leq n_i-1} \psi(e_{ij})\psi(e_{ij+1}). \qquad (10.8)$$

For the tuning parameter, Fan et al. (2012) proposed a robust generalized CV to choose λ for reducing the impact of outliers. Define $\mathrm{RSS}_R(\lambda)$ as the summation of squares of the robustified residuals, that is, $\mathrm{RSS}_R(\lambda) = \sum_{i=1}^{N} \left\| W_i \psi(A_i^{-1/2}(Y_i - \mu_i(\hat{\beta}_\lambda))) \right\|^2$, where $\hat{\beta}_\lambda$ is the solution of the penalized robust estimating equation (10.5) when λ is fixed. Select the penalty parameter λ by minimizing the robustified GCV statistic:

$$\mathrm{GCV}_R(\lambda) = \frac{\mathrm{RSS}_R(\lambda)/n}{\{1 - d(\lambda)/n\}^2}, \qquad (10.9)$$

where $d(\lambda) = \mathrm{tr}\{[D(\hat{\beta}) + NE(\hat{\beta}_\lambda)]^{-1}D^\mathrm{T}(\hat{\beta}_\lambda)\}$ is the effective number of parameters. Then $\lambda_{\mathrm{opt}} = \mathrm{argmin}_\lambda \mathrm{GCV}_R$ is chosen, and the corresponding $\hat{\beta}_\lambda$ is the robust penalized estimator of β.

The specific algorithm is given as follows.

- Step 1. We can use the estimator $\hat{\beta}_0$ obtained from GEE with an independence working matrix as an initial estimator of β.

- Step 2. Update the correlation parameter α and scale parameter ϕ using (10.6):

- Step 3. For a given tuning parameter λ, utilize the MM algorithm to solve Equation (10.4). Let $\hat{\beta}^{(k)}$ be the kth iterate value for a fixed λ, then the iterative MM algorithm is written:

$$\hat{\beta}^{(k+1)} = \hat{\beta}^{(k)} - [D(\hat{\beta}^{(k)}) - NE(\hat{\beta}^{(k)})]^{-1}[U_R(\hat{\beta}) - NE(\hat{\beta}^{(k)}) \times \hat{\beta}^{(k)}], \qquad (10.10)$$

where $D(\beta) = \partial U_R(\beta)/\partial\beta$. The MM algorithm continues until successive iterates values are less than a user-defined threshold.

- Step 4. Choose λ using the GCV estimator.

10.3   Smooth-threshold Method

Ueki (2009) developed smooth-threshold estimating equations that can automatically eliminate irrelevant parameters by setting them as zero, which can be extended to the marginal model. Consider the following equations:

$$(I_p - \Delta)U_R(\beta) + \Delta\beta = 0_p, \qquad (10.11)$$

where I_p is a p-dimensional identity matrix, and Δ is a diagonal matrix with diagonal elements $(\hat{\delta}_1, \cdots, \hat{\delta}_p)$, in which $\hat{\delta}_j = \min\left\{1, \lambda^\star / \left|\hat{\beta}_j^{(0)}\right|^{(1+\tau)}\right\}$ with a $\sqrt{N}$-consistent estimator $\hat{\beta}_j^{(0)}$ $(j = 1, \cdots, p)$ which can be obtained by solving equation (10.4). Parameter τ and λ are tuning parameters. If λ takes a large value or τ takes a small value, then $\hat{\delta}_j = 1$, and hence the corresponding parameter is shrunk to zero and the corresponding covariate X_{ij} can be removed from the model. Before solving the equation, we need to specify the parameters τ and λ. According to Zou (2006), parameter τ can be selected among $(0.5, 1, 2)$. The parameter $\lambda^\star$ is a tuning parameter and controls the sparsity of parameters. To solve equation (10.11), $\lambda^\star$ should be specified. The following criterion function can be used to select a proper tuning parameter $\lambda^\star$:

$$\text{RBIC}(\lambda^\star) = \sum_{i=1}^{N} h_i^T(\mu_i(\hat{\beta}_{\lambda\star})) R_i^{-1} h_i^T(\mu_i(\hat{\beta}_{\lambda\star})) + df_{\lambda\star} \log N, \tag{10.12}$$

where $\hat{\beta}_{\lambda\star}$ is an estimator of β for a given $\lambda^\star$, and $df_{\lambda\star} = \sum_{i=1}^{p} I(\hat{\delta}_j \neq 1)$ is the number of nonzero elements of the estimators. The optimal tuning parameter $\lambda^\star_{opt} = \min_{\lambda\star} \text{RBIC}(\lambda^\star)$. The procedure for solving equation (10.11) follows.

- Step 1. Give an initial estimate $\hat{\beta}^{(0)}$. For example, the estimate obtained via the generalized estimating equations with an independence working matrix can be used as the initial estimate.

- Step 2. Estimate the scale parameter $\hat{\phi}$ using (10.6) with the current estimate $\hat{\beta}^{(k)}$. For a given working correlation matrix $R_i(\alpha)$, estimate the correlation parameter using (10.7) and (10.8) for exchangeable and AR(1), respectively, and obtain

$$V_i\left\{\mu_i\left(\hat{\beta}^{(k)}\right), \hat{\phi}^{(k)}\right\} = R_i(\hat{\alpha}^{(k)}) \hat{A}_i^{1/2}\left(\hat{\beta}^{(k)}, \hat{\phi}^{(k)}\right).$$

Step 3. For a given λ, update the estimate of β via the following iterative formula:

$$\hat{\beta}^{(k+1)} = \hat{\beta}^{(k)} - \left\{\left(\sum_{i=1}^{n} D_i^T \Omega_i(\mu_i(\beta)) D_i + \hat{G}\right)^{-1} \left(U_n(\beta) + \hat{G} \cdot \beta\right)\right\}\Bigg|_{\hat{\beta}^{(k)}}, \tag{10.13}$$

where $\hat{G} = \left(I_p - \hat{\Delta}\right)^{-1}\hat{\Delta}$, and $\Omega(\mu_i(\beta)) = V_i^{-1}(\mu_i(\beta))\Gamma_i(\mu_i(\beta))$, in which

$$\Gamma_i(\mu_i(\beta)) = E\left[\dot{h}_i^b(\mu_i(\beta))\right] = \mathbf{E}\left[\partial h_i^b(\mu_i(\beta))/\partial\mu_i\right]\Big|_{\mu_i = \mu_i(\beta)},$$

Step 4. Repeat Steps 2–3 until the algorithm converges.

According to the iterative algorithm mentioned above, we obtain a sandwich formula to estimate the asymptotic covariance matrix of $\hat{\beta}_\lambda$:

$$\text{Cov}(\hat{\beta}_\lambda) \approx \left[\hat{\Sigma}_n\left(\mu_i(\hat{\beta}_\lambda)\right)\right]^{-1} \hat{\mathbf{H}}_n\left(\mu_i(\hat{\beta}_\lambda)\right)\left[\hat{\Sigma}_n\left(\mu_i(\hat{\beta}_\lambda)\right)\right]^{-1}, \tag{10.14}$$

where

$$\hat{\mathbf{H}}_n\left(\mu_i(\hat{\beta}_\lambda)\right) = \sum_{i=1}^{n} D_i^T V_i^{-1}\left(\mu_i(\hat{\beta}_\lambda)\right)\left[h_i^b\left(\mu_i(\hat{\beta}_\lambda)\right)\left\{h_i^b\left(\mu_i(\hat{\beta}_\lambda)\right)\right\}^T\right]V_i^{-1}\left(\mu_i(\hat{\beta}_\lambda)\right)D_i^T,$$

and

$$\hat{\Sigma}_n\left(\mu_i(\hat{\beta}_\lambda)\right) = \sum_{i=1}^n D_i^{\mathrm{T}} V_i^{-1}\left(\mu_i(\hat{\beta}_\lambda)\right)\Gamma_i\left(\mu_i(\hat{\beta}_\lambda)\right)D_i.$$

The tuning parameter b can be selected by minimizing the determinant of the covariance matrix of $\hat{\beta}_\lambda$, that is $b_{opt} = \operatorname{argmin}_b |\operatorname{Cov}(\hat{\beta}_\lambda)|$.

10.4 Yeast Data Study

In this section, we applied the variable selection methods to analyze a yeast dataset. A yeast cell-cycle gene expression dataset was collected in the CDC15 experiment where genome-wide mRNA levels of 6178 yeast open reading frames (ORFs) in a two cell-cycle period were measured at M/G1-G1-S-G2-M stages (Spellman et al., 1998). To better understand the phenomenon underlying cell-cycle process, it is important to identify transcription factors (TFs) that regulate the gene expression levels of cell cycle-regulated genes. In this section, we use a subset of 283 cell-cycled-regularized genes observed over 4 time points at G1 stage, which is available in the R package *PGEE* (Inan et al., 2016).

The response variable Y_{ik} is the log-transformed gene expression level of gene i measured at time point k. The covariates x_{ij}, $j = 1, \cdots, 96$, are the standardized binding probabilities of a total of 96 TFs obtained from a mixture model approach of Wang et al. (2007) based on the ChIP data of Lee et al. (2002). The full model considered is,

$$Y_{ik} = \beta_0 + \beta_1 t_{ik} + \sum_{j=1}^{96} \beta_j x_{ij} + \epsilon_{ik},$$

where t_{ik} denotes time, and $x_{i1}, \cdots, x_{i96}$ are standardized to have mean zero and unit variance.

Figure 10.4 shows the change of the log-transformed gene expressive level with time for the first twenty genes. The histogram (see Figure 10.4) indicates the distribution of log-transformed gene expressive level is symmetric. The boxplot (see Figure 10.3) indicates that log-transformed gene expression levels may contain many underlying outliers. We utilize the penalized methods to identify the important TFs. Table 10.1 presents the number of TFs selected for the G1-stage yeast cell-cycle process using the penalized GEE, penalized Huber (c=2) and penalized exponential squared loss with SCAD penalty under three correlation structures (independence, exchangeable and AR(1)). The results indicate that these three methods tend to select more TFs under independence correlation structure. Table 10.2 lists the selected TFs using the three penalized methods under independence, exchangeable and AR(1) correlation structures. The overlaps of selected TFs using the three methods can be treated as important TFs. It would be of great interest to further study other controversial TFs and confirm their biological properties using the genome-wide binding method.

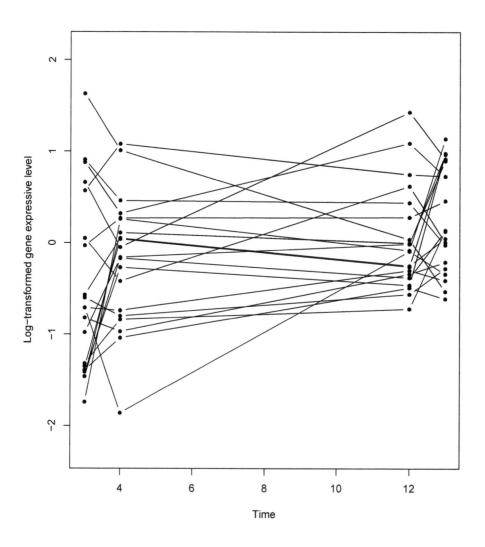

Figure 10.1 *Plot of the log-transformed gene expressive level with time for the first twenty genes.*

10.5   Further Reading

The aforementioned methods in Sections §10.1 and §10.2.2 can only deal with the data that the number of covariates p is no more than the number of subjects N. Therefore, when p grows at an exponential rate of N, that is, the longitudinal data is

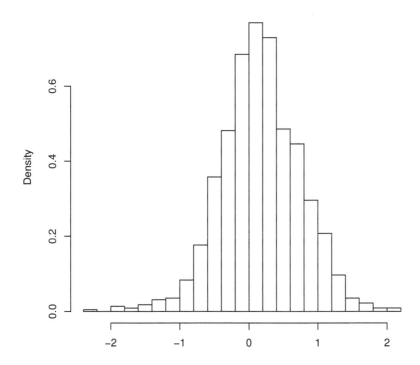

Figure 10.2 *Histogram of the log-transformed gene expressive level in the yeast cell-cycle process.*

Table 10.1 *The number of TFs selected for the G1-stage yeast cell-cycle process with penalized GEE (PGEE), penalized Exponential squared loss (PEXPO), and penalized Huber loss (PHUBER) with SCAD penalty.*

Correlation	PGEE	PHUBER	PEXPO
Independence	18	16	15
Exchangeable	13	12	11
AR(1)	14	14	11

ultra-high dimensional, new methods need to be developed. Ultra-high dimensional longitudinal data are increasingly common and the analysis is challenging both theoretically and methodologically.

Cheng et al. (2014) and Fan et al. (2014) studied the varying coefficient model under ultra-high dimensional longitudinal data. They proposed first reducing the num-

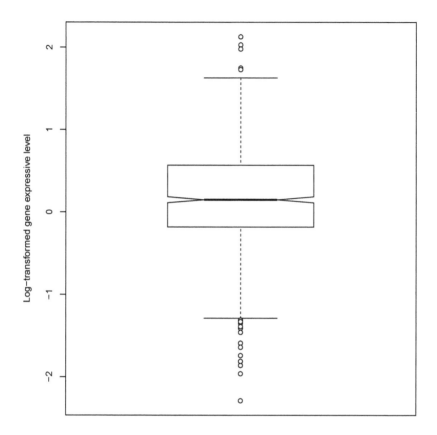

Figure 10.3 *Boxplot of the log-transformed gene expressive level in the yeast cell-cycle process.*

ber of covariates to a moderate order by employing a screening procedure proposed by Fan and Lv (2008), and then identified both the varying and constant coefficients using a group SCAD. Their method is based on B-spline marginal models and working independence assumption, and the correlations are ignored.

Xu et al. (2014) proposed a novel GEE-based screening procedure, which only pertains to the specifications of the first two marginal moments and a working correlation structure. Different from existing methods that either fit separate marginal models or compute pairwise correlation measures, their method merely involves making a single evaluation of estimating functions and thus is extremely computationally efficient. The new method is robust against the misspecification of corre-

Table 10.2 *List of selected TFs for the G1-stage yeast cell-cycle process with penalized GEE (PGEE), penalized Exponential squared loss (PEXPO), and penalized Huber loss (PHUBER) with SCAD penalty.*

	PGEE							
Independence	ABF1	FKH1	FKH2	GAT3	GCR2	MBP1	MSN4	NDD1
	PHD1	RGM1	RLM1	SMP1	SRD1	STB1	SWI4	SWI6
Exchangeable	FKH1	FKH2	GAT3	MBP1	NDD1	PHD1	RGM1	SMP1
	STB1	SWI4	SWI6					
AR(1)	FKH1	FKH2	GAT3	MBP1	MSN4	NDD1	PHD1	RGM1
	SMP1	STB1	SWI4	SWI6				
	PHUBER							
Independence	FKH1	FKH2	GAT3	GCR2	IXR1	MBP1	NDD1	NRG1
	PDR1	ROX1	SRD1	STB1	STP1	SWI4	SWI6	YAP5
Exchangeable	FKH2	GAT3	GCR2	MBP1	NDD1	PDR1	SRD1	STB1
	STP1	SWI4	SWI6	YAP5				
AR(1)	FKH1	FKH2	GAT3	GCR2	MBP1	NDD1	NRG1	PDR1
	SRD1	STB1	STP1	SWI4	SWI6	YAP5		
	PEXPO							
Independence	ABF1	FKH1	FKH2	GAT3	GCR2	MBP1	MSN4	NDD1
	PDR1	PHD1	RGM1	RLM1	SRD1	SWI4	SWI6	
Exchangeable	ABF1	FKH2	GAT3	GCR2	MBP1	NDD1	RLM1	SRD1
	SWI4	SWI6	YAP5					
AR(1)	ABF1	FKH2	GAT3	GCR2	MBP1	NDD1	RLM1	SRD1
	SWI4	SWI6	YAP5					

lation structures. Liu (2016) studied longitudinal partially linear models with ultra high-dimensional covariates, and provided a two-stage variable selection procedure that consists of a quick screening stage and a post-screening refining stage. The proposed approach is based on the partial residual method for dealing with the nonparametric baseline function.

Li et al. (2018) proposed using a robust conditional quantile correlation or conditional distribution correlation screening procedures to reduce dimension of the covariates to a moderate order and then utilize the kernel smoothing technique to estimate the population conditional quantile correlation and population conditional distribution correlation for the varying coefficient models in ultra high-dimensional data analysis.

These studies are mainly about continuous longitudinal data, and there are a few studies on discrete longitudinal data. In addition, when the variables are filtered in the ultra-high-dimensional longitudinal data, the screening methods for independent ultra-high-dimensional data cannot be directly applied to the ultra-high-dimensional longitudinal data. Zhu et al. (2017) proposed that projection correlation for random vector correlation, which provides a new idea for the study of correlation between vectors. However, there are still many problems need to be resolved in the correlation between vectors.

Bibliography

P. S. Albert, and L. M. McShane. A generalized estimating equations approach for spatially correlated binary data: Applications to the analysis of neuroimaging data. *Biometrics*, 51:627–638, 1995.

P. S. Albert. A transitional model for longitudinal binary data subject to nonignorable missing data. *Biometrics*, 56(2):602–608, 2000.

P. S. Albert, and D. A. Follmann. A random effects transition model for longitudinal binary data with informative missingness. *Statistica Neerlandica*, 57(1):100–111, 2003.

P. M. E. Altham. Two generalizations of the binomial distribution. *Journal of the Royal Statistical Society. Series C (Applied Statistics)*, 27(2):162–167, 1978.

T. W. Anderson, and J. B. Taylor. Strong consistency of least squares estimates in normal linear regression. *The Annals of Statistics*, 4(4):788–790, 1976.

F. J. Anscombe. The statistical analysis of insect counts based on the negative binomial distribution. *Biometrics*, 5(2):165–173, 1949.

R. K. Barnwal, and S. R. Paul. Analysis of one-way layout of count data with negative binomial variation. *Biometrika*, 75(2):215–222, 1988.

V. P. Bhapkar. On a measure of efficiency of an estimating equation. *Sankhyā: The Indian Journal of Statistics, Series A*, 34(4):467–472, 1972.

C. I. Bliss, and R. A. Fisher. Fitting the negative binomial distribution to biological data. *Biometrics*, 9(2):176–200, 1953.

D. Böhning, E. Dietz, P. Schlattmann, L. Mendonca, and U. Kirchner. The zero-inflated poisson model and the decayed, missing and filled teeth index in dental epidemiology. *Journal of the Royal Statistical Society: Series A (Statistics in Society)*, 162(2):195–209, 1999.

N. E. Breslow. Extra-poisson variation in log-linear models. *Journal of the Royal Statistical Society: Series C (Applied Statistics)*, 33(1):38–44, 1984.

B. M. Brown, and Y-G. Wang. Standard errors and covariance matrices for smoothed rank estimators. *Biometrika*, 92:149–158, 2005.

A. Callens, Y. Wang, L. Y. Fu, and B. Liquet. Robust estimation procedure for autoregressive models with heterogeneity. *Environmental Modeling and Assessment*, 2020. doi: 10.1007/s10666-020-09730-w.

V. J. Carey, and Y-G. Wang. Working covariance model selection for generalized estimating equations. *Statistics in Medicine*, 30:3117–3124, 2011.

V. J. Carey, S. L. Zeger, and P. Diggle. Modelling multivariate binary data with alternating logistic regressions. *Biometrika*, 80:517–526, 1993.

J. P. Carpenter, S. Pocock, and C. J. Lamm. Coping with missing data in clinical trials: a model-based approach applied to asthma trials. *Statistics in Medicine*, 21:43–66, 2002.

R. J. Carroll, and D. Ruppert. Robust estimation in heteroscedastic linear models. *The Annals of Statistics*, 10(2):429–441, 1982.

G. Casella, and R. L. Berger. *Statistical inference*, volume 2. Duxbury Pacific Grove, CA, 2002.

G. Casella, and E. I. George. Explaining the gibbs sampler. *The American Statistician*, 46(3):167–174, 1992.

R. N. Chaganty. An alternative approach to the analysis of longitudinal data via generalized estimating equations. *Journal of Statistical Planning and Inference*, 63:39–54, 1997.

R. N. Chaganty, and J. Shults. On eliminating the asymptotic bias in the quasi-least squares estimate of the correlatiopn parameter. *Journal of Statistical Planning and Inference*, 76:145–161, 1999.

G. Chamberlain. Quantile regression, censoring and the structure of wages. *In Proceedings of the Sixth World Congress of the Econometrics Society (eds. C. Sims and J.J. Laffont)*, 2:171–209, 1994.

J. Chen, and N. A. Lazar. Selection of working correlation structure in generalized estimating equations via empirical likelihood. *Journal of Computational and Graphical Statistics*, 21:18–41, 2012.

J. Chen, S. Hubbard, and Y. Rubin. Estimating the hydraulic conductivity at the south oyster site from geophysical tomographic data using bayesian techniques based on the normal linear regression model. *Water Resources Research*, 37(6): 1603–1613, 2001.

L. Chen, L. J. Wei, and M. I. Parzen. Quantile regression for correlated observations. *Proceedings of the Second Seattle Symposium in Biostatistics: Analysis of Correlated Data*, 179:51–70, 2004.

M.-H. Chen, and J. G. Ibrahim. Maximum likelihood methods for cure rate models with missing covariates. *Biometrics*, 57:43–52, 2002.

M-Y. Cheng, T. Honda, J. L. Li, and H. Peng. Nonparametric independence screening and structure identification for ultra-high dimensional longitudinal data. *The Annals of Statistics*, 42:1819–1849, 2014.

S. J. Clark, and J. N. Perry. Estimation of the negative binomial parameter κ by maximum quasi-likelihood. *Biometrics*, 45(1):309–316, 1989.

D. R. Cox. Regression models and life-tables. *Journal of the Royal Statistical Society, Series B*, 34(2):187—220, 1972.

D. R. Cox, and N. Reid. Parameter orthogonality and approximate conditional inference. *Journal of the Royal Statistical Society. Series B*, 1–39, 1987.

D. R. Cox, and E. J. Snell. *Analysis of Binary Data, Second Edition*. Chapman & Hall/CRC Monographs on Statistics & Applied Probability. Taylor & Francis, 1989.

M. Crowder. Gaussian estimation for correlated binomial data. *Journal of the Royal Statistical Society, Series B*, 47:229–237, 1985.

M. Crowder. On the use of a working correlation matrix in using generalised linear models for repeated measures. *Biometrika*, 82:407–410, 1995.

M. Crowder. On repeated measures analysis with misspecified covariance structure. *Journal of Royal Statistical Society, Statistical Methodology, Series B*, 63(1): 55–62, 2001.

S. Datta, and A. Satten, Glen. Rank-sum tests for clustered data. *Journal of the American Statistical Association*, 100(471):908–915, 2005.

M. Davidian, and R. J. Carroll. Variance function estimation. *Journal of the American Statistical Association*, 82(400):1079–1091, 1987.

M. Davidian, and D. M. Giltinan. *Nonlinear Models for Repeated Measurement Data*. Chapman & Hall, London, 1995.

C. S. Davis. Semi-parametric and non-parametric methods for the analysis of repeated measurements with applications to clinical trials. *Statistics in Medicine*, 10(12):1959–1980, 1991.

I. Deltour, S. Richardson, and J.-Yves L. Hesran. Stochastic algorithms for markov models estimation with intermittent missing data. *Biometrics*, 55(2):565–573, 1999.

A. P. Dempster, N. M. Laird, and D. B. Rubin. Maximum likelihood from incomplete data via the em algorithm. *Journal of the Royal Statistical Society. Series B*, 39: 1–38, 1977.

A. M. Dempster. Covariance selection. *Biometrics*, 28:157–175, 1972.

A. P. Dempster, C. M. Patel, and A. J. Selwyn, M. R., and Roth. Statistical and computational aspects of mixed model analysis. *Journal of the Royal Statistical Society. Series C (Applied Statistics)*, 33(2):203–214, 1984.

D. Deng, and S. R. Paul. Score tests for zero-inflation and over-dispersion in generalized linear models. *Statistica Sinica*, 15:257–276, 2005.

P. Diggle, and M. G. Kenward. Informative drop-out in longitudinal data analysis (with discussion). *Applied Statistics*, 43:49–93, 1994.

P. J. Diggle. An approach to the analysis of repeated measurements. *Biometrics*, 44 (4):959–971, 1988.

P. J. Diggle, K-Y. Liang, and S. L. Zeger. *Analysis of Longitudinal Data*. Clarendon Press, 1994.

P. J. Diggle, K-Y. Liang, and S. L. Zeger. *Analysis of Longitudinal Data*. Oxford University Press, Oxford, 2002.

A. Donner. A review of inference procedures for the intraclass correlation coefficient in the one-way random effects model. *International Statistical Review/Revue Internationale de Statistique*, 54(1):67–82, 1986.

A. Donner, and S. Bull. Inferences concerning a common intraclass correlation coefficient. *Biometrics*, 39(3):771–775, 1983.

A. Donner, and J. J. Koval. The estimation of intraclass correlation in the analysis of family data. *Biometrics*, 36(1):19–25, 1980.

T. M. Durairajan. Optimal estimating function for non-orthogonal model. *Journal of Statistical Planning Inference*, 33:381–384, 1992.

F. Eicker. Asymptotic normality and consistency of the least squares estimators for families of linear regressions. *The Annals of Mathematical Statistics*, 34(2):447–456, 1963.

A. Ekholm, and C. Skinner. The muscatine children's obesity data reanalysed using pattern mixture models. *Journal of the Royal Statistical Society: Series C (Applied Statistics)*, 47(2):251–263, 1998.

J. Engel. Models for response data showing extra-poisson variation. *Statistica Neerlandica*, 38(3):159–167, 1984.

J. Fan, and R. Li. Variable selection via nonconcave penalized likelihood and its oracle properties. *Journal of the American Statistical Association*, 96:1348–1360, 2001.

J. Fan, and J. Lv. Sure independence screening for ultrahigh dimensional feature space. *Journal of the Royal Statistical Society, Ser. B*, 70:849–911, 2008.

J. Fan, Y. Ma, and W. Dai. Nonparametric independence screening in sparse ultra-high-dimensional varying coefficient models. *Journal of the American Statistical Association*, 109:1270–1284, 2014.

Y. L. Fan, G. Y. Qin, and Z. Y. Zhu. Variable selection in robust regression models for longitudinal data. *Journal of Multivariate Analysis*, 109:156–167, 2012.

R. A. Fisher. *Statistical methods for research workers*. Genesis Publishing Pvt Ltd, 1925.

G. M. Fitzmaurice. A caveat concerning independence estimating equations with multivariate binary data. *Biometrics*, 51:309–317, 1995.

G. M. Fitzmaurice, N. M. Laird, and J. H. Ware. Applied Longitudinal Analysis, 2nd Edition, Wiley.

G. M. Fitzmaurice, N. M. Laird, and A. G. Rotnitzky. Regression models for discrete longitudinal responses (disc: P300-309). *Statistical Science*, 8:284–299, 1993.

G. M. Fitzmaurice, S. R. Lipsitz, Gelber R. Ibrahim JG, and Lipshultz. S. Estimation in regression models for longitudinal binary data with outcome-dependent follow-up. *Biostatistics*, 7:469–465, 2006.

L. Y. Fu, and Y-G. Wang. Efficient estimation for rank-based regression with clustered data. *Biometrics*, 68:1074–1082, 2012a.

L. Y. Fu, Y-G. Wang, and Z. D. Bai. Rank regression for analysis of clustered data: A natural induced smoothing approach. *Computational Statistics and Data Analysis*, 54:1036–1050, 2010.

L. Y. Fu, Y-G. Wang, and M. Zhu. A gaussian pseudolikelihood approach for quantile regression with repeated measurements. *Computational Statistics and Data Analysis*, 84:41–53, 2015.

L. Y. Fu, Y-G. Wang, and F. Cai. A working likelihood approach for robust regression. *Statistical Methods in Medical Research*, 29(12):3641–3652, 2020.

L. Y. Fu, and Y-G. Wang. Quantile regression for longitudinal data with a working correlation model. *Computational Statistics and Data Analysis*, 56:2526–2538, 2012b.

L. Y. Fu, and Y-G. Wang. Efficient parameter estimation via gaussian copulas for quantile regression with longitudinal data. *Journal of Multivariate Analysis*, 143:492–502, 2016.

K. W. Fung, Z. Y. Zhu, B. C. Wei, and X. M. He. Infference diagnostics and outlier tests for semiparametric mixed models. *Journal of Royal Statistical Society, Series B.*, 64:565–579, 2002.

J. Geweke. Exact inference in the inequality constrained normal linear regression model. *Journal of Applied Econometrics*, 1(2):127–141, 1986.

W. R. Gilks, and P. Wild. Adaptive rejection sampling for gibbs sampling. *Applied Statistics*, 41:337–348, 1992.

V. P. Godambe, and M. E. Thompson. An extension of quasi-likelihood estimation. *Statistical Planning and Inference*, 22:137–152, 1989a.

V. P. Godambe. An optimum property of regular maximum likelihood estimation (Ack: V32 p1343). *The Annals of Mathematical Statistics*, 31:1208–1212, 1960.

V. P. Godambe, and C. C. Heyde. Quasi-likelihood and optimal estimation. *International Statistical Review*, 55:231–244, 1987.

V. P. Godambe, and Mary Elinore Thompson. An extension of quasi-likelihood estimation. *Journal of Statistical Planning and Inference*, 22(2):137–152, 1989b.

D. Griffin, and R. Gonzalez. Correlational analysis of dyad-level data in the exchangeable case. *Psychological Bulletin*, 118(3):430, 1995.

D. Hall, and T. A. Severini. Extended generalized estimating equations for clustered data. *Journal Americian Statistican and Association*, 93:1365–1375, 1998.

L. P. Hansen. Large sample properties of generalized method of moments estimators. *Econometrica*, 50(4):1029–1054, 1982.

J. Hardin, and J. Hilbe. *Generalized Estimating Equations*. Chapman & Hall, CRC, 2012.

D. A. Harville, and D. R. Jeske. Mean squared error of estimation or prediction under a general linear model. *Journal of the American Statistical Association*, 87(419):724–731, 1992.

J. K. Haseman, and L. L. Kupper. Analysis of dichotomous response data from certain toxicological experiments. *Biometrics*, 35(1):281–293, 1979.

T. P. Hettmansperger. *Statistical Inference Based on Ranks*. New York: John Wiley and Sons, 1984.

C. C. Heyde. *Statistical Data Analysis and Inference*. Amsterdam: Elsevier, 1987.

C. C. Heyde. *Quasi-Likelihood and Its Application: A General Approach to Optimal Parameter Estimation*. New York: Springer, 1997.

L. Y. Hin, and Y-G. Wang. Working-correlation-structure identification in generalized estimating equations. *Statistics in Medicine*, 28(4):642–658, 2009.

D. D. Ho, A. U. Neumann, A. S. Perelson, W. Chen, J. M. Leonard, and M. Markowitz. Rapid dynamics in human immunodeficiency virus type 1 turnover of plasma virions and cd4 lymphocytes in hiv-1 infection. *Nature*, 373:123–126, 1995.

E. B. Hoffman, P. K. Sen, and C. R. Weinberg. Within-cluster resampling. *Biometrika*, 88(4):1121–1134, 2001.

L. Hua, and Y. Zhang. Spline-based semiparametric projected generalized estimating equation method for panel count data. *Biostatistics*, 13(3):440–454, 2012.

A. Huang, and P. J. Rathouz. Proportional likelihood ratio models for mean regression. *Biometrika*, 99:223–229, 2012.

P. J. Huber. The behavior of maximum likelihood estimates under nonstandard conditions. In *Proceedings of the Fifth Berkeley Symposium on Mathematical Statistics and Probability, Volume 1: Statistics*, 221–233. University of California Press, 1967. https://projecteuclid.org/euclid.bsmsp/1200512988.

D. R. Hunter, and R. Li. Variable selection using mm algorithms. *The Annals of Statistic*, 33:1617–1642, 2005.

J. G. Ibrahim. Incomplete data in generalized linear models. *Journal of the American Statistical Association*, 85:765–769, 1990.

J. G. Ibrahim, and S. R. Lipsit. Parameter estimation from incomplete data in binomial regression when the missing data mechanism is nonignorable. *Biometrics*, 52:1071–1078, 1996.

J. G. Ibrahim, and G. Molenberghs. Missing data methods in longitudinal studies: a review. *Test*, 18:1–43, 2009.

J. G. Ibrahim, M. H. Chen, and S. R. Lipsitz. Missing responses in generalized linear mixed models when the missing data mechanismis nonignorable. *Biometrika*, 88:551–564, 2001.

J. G. Ibrahim, M. H. Chen, S. R. Ipsitiz, and A. H. Herring. Missing-data methods for generalized linear models: A comparative review. *Journal of the American Statistical Association*, 100:332–346, 2005.

J. G. Ibrahim. Incomplete data in generalized linear models. *Journal of the American Statistical Association*, 85(411):765–769, 1990.

J. G. Ibrahim, and S. R. Lipsitz. Parameter estimation from incomplete data in binomial regression when the missing data mechanism is nonignorable. *Biometrics*, 52(3):1071–1078, 1996.

G. Inan, and L. Wang. Pgee: An r package for analysis of longitudinal data with high-dimensional covariates. *The R Journal*, 9:393–402, 06 2017. doi: 10.32614/RJ-2017-030.

G. Inan, J. H. Zhou, and L. Wang. Pgee: Penalized generalized estimating equations in high-dimension. https://cran.r-project.org/package=pgee. *R package version 1.5*, https://CRAN.R-project.org/package=PGEE, 2016.

R. I. Jennrich, and M. D. Schluchter. Unbalanced repeated-measures models with structured covariance matrices. *Biometrics*, 42:805–820, 1986.

J. M. Jiang. *Linear and Generalized Linear Mixed Models and Their Applications*. Springer, New York, 2007.

J. M. Jiang, and T. Nguyen. *The Fence Methods*. World Scientific, 2015.

B. Jørgensen. The Theory of Dispersion Models. London: Chapman and Hall, 1997.

S. H. Jung. Quasi-likelihood for median regression models. *Journal of the American Statistical Association*, 91:251–257, 1996.

S. H. Jung, and Z. Ying. Rank-based regression with repeated measurements data. *Biometrika*, 90:732–740, 2003.

M. C. K. Tweedie. *An index which distinguishes between some important exponential families*. A4, 22 pp. Pre-print issued for the Golden Jubliee Conference at the Indian Statistical Institute, Calcutta, 1981.

G. Kauerman and R.J. Carroll. A note on the efficiency of sandwich covariance matrix estimation. *Journal of the American Statistical Association*, 96:1387–1396, 2001.

B. C. Kelly. Some aspects of measurement error in linear regression of astronomical data. *The Astrophysical Journal*, 665(2):1489, 2007.

A. I. Khuri, and H. Sahai. Variance components analysis: a selective literature survey. *International Statistical Review/Revue Internationale de Statistique*, 53(3):279–300, 1985.

J. C. Kleinman. Proportions with extraneous variance: single and independent samples. *Journal of the American Statistical Association*, 68(341):46–54, 1973.

R. Koenker, and G. Jr. Bassett. Regression quantiles. *Econometrica*, 84:33–50, 1978.

R. Koenker, and V. D'Orey. Computing regression quantiles. *Applied Statistics*, 36: 383–393, 1987.

E. L. Korn, and A. S. Whittemore. Methods for analyzing panel studies of acute health effects of air pollution. *Biometrics*, 35:795–802, 1979.

L. L. Kupper, and J. K. Haseman. The use of a correlated binomial model for the analysis of certain toxicological experiments. *Biometrics*, 34(1):69–76, 1978.

N. M. Laird, and J. H. Ware. Random-effects models for longitudinal data. *Biometrics*, 38:963–974, 1982.

N. Lange, L. Ryan, L. Billard, D. Brillinger, L. Conquest, and J. (eds.) Greenhouse. Case Studies in Biometry. New York: Wiley-Interscience, 1994.

J. F. Lawless. Regression methods for poisson process data. *Journal of the American Statistical Association*, 82(399):808–815, 1987.

E. W. Lee, and M. Y. Kim. The analysis of correlated panel data using a continuous-time markov model. *Biometrics*, 54(4):1638–1644, 1998.

J. C. Lee. Prediction and estimation of growth curves with special covariance structures. *Journal of the American Statistical Association*, 83:432–440, 1988.

Y. Lee, and J. A. Nelder. Hierarchical generalised linear models: A synthesis of generalised linear models, random-effect models and structured dispersions. *Biometrika*, 88:987–1006, 2001.

C. L. Leng, and H. P. Zhang. Smoothing combined estimating equations in quantile regression for longitudinal data. *Statistics and Computing*, 24:123–136, 2014.

G. Li, L. Zhu, L. Xue, and S. Feng. Empirical likelihood inference in partially linear single-index models for longitudinal data. *Journal of Multivariate Analysis*, 101 (3):718–732, 2010.

X. J. Li, X. J. Ma, and J. X. Zhang. Conditional quantile correlation screening procedure for ultrahigh-dimensional varying coefficient models. *Journal of Statistical Planning and Inference*, 197:69–92, 2018.

K-Y. Liang, and S. L. Zeger. Lonagitudinal data analysis using generalized linear models. *Biometrika*, 73:13–22, 1986.

K-Y. Liang, S. L. Zeger, and B. Qaqish. Multivariate regression analyses for categorical data (disc: P24-40). *Journal of the Royal Statistical Society, Series B*, 54: 3–24, 1992.

D. Y. Lin, and Z. Ying. Semiparametric and nonparametric regression analysis of longitudinal data. *Journal of the American Statistical Association*, 96(453):103–126, 2001.

X. Lin, and R. J. Carroll. Semiparametric estimation in general repeated measures problems. *Journal of the Royal Statistical Society: Series B*, 68(1):69–88, 2006.

M. J. Lindstrom, and D. M. Bates. Newton—raphson and em algorithms for linear mixed-effects models for repeated-measures data. *Journal of the American Statistical Association*, 83(404):1014–1022, 1988.

S. R. Lipsitz, and G. M. Fitzmaurice. Estimating equations for measures of association between repeated binary responses. *Biometrics*, 52(3):903–912, 1996.

S. R. Lipsitz, N. M. Laird, and D. P. Harrington. Finding the design matrix for the marginal homogeneity model. *Biometrika*, 77:353–358, 1990.

S. R. Lipsitz, N. M. Laird, and D. P. Harrington. Generalized estimating equations for correlated binary data: Using the odds ratio as a measure of association. *Biometrika*, 78:153–160, 1991.

S. R. Lipsitz, K. n Kim, and L. Zhao. Analysis of repeated categorical data using generalized estimating equations. *Statistics in Medicine*, 13:1149–1163, 1994.

S. R. Lipsitz, J. G. Ibrahim, and G. Molenburgs. Using a box-cox transformation in the analysis of longitudinal data with incomplete responses. *Applied Statistics*, 49:287–296, 2000.

S. R. Lipsitz, G. M. Fitzmaurice, J. G. Ibrahim, and R. Gelber. Parameter estimation in longitudinal studies with outcome-dependent follow-up. *Biometrics*, 58:621–630, 2002.

R. J. A. Little. Pattern-mixture models for multivariate incomplete data. *Journal of the American Statistical Association*, 88:125–134, 1993.

R. J. A. Little. Modeling the drop-out mechanism in repeated-measures studies. *Jornal of the American Statistical Association*, 90:1113–1121, 1995.

R. J. A. Little, and D. B. Rubin. *Statistical Analysis with Missing Data, 2nd Edition*. Wiley, 1987.

J. Liu. Feature screening and variable selection for partially linear models with ultrahigh-dimensional longitudinal data. *Neurocomputing*, 195:202–210, 2016.

K-J. Lui, Mayer J.A., and L. Eckhardt. Confidence intervals for the risk ratio under cluster sampling based on the beta-binomial model. *Statistics in Medicine*, 19 (21):2933–2942, 2000.

J. Lv, H. Yang, and C. H. Guo. An efficient and robust variable selection method for longitudinal generalized linear models. *Computational Statistics and Data Analysis*, 82:74–88, 2015.

M. Gosho, C. Hamada, and I. Yoshimura. Selection of working correlation structure in weighted generalized estimating equation method for incomplete longitudinal data. *Communications in Statistics - Simulation and Computation*, 43:62–81, 2014.

T. Maiti, and V. Pradhan. Bias reduction and a solution for separation of logistic regression with missing covariates. *Biometrics*, 65(4):1262–1269, 2009.

L. A. Mancl, and T. A. DeRouen. A covariance estimator for gee with improced samll-sample properties. *Biometrice*, 57:126–134, 2001.

L. A. Mancl, and B. G. Leroux. Efficiency of regression estimates for clustered data. *Biometrics*, 52:500–511, 1996.

K. G. Manton, M. A. Woodbury, and E. Stallard. A variance components approach to categorical data models with heterogenous cell populations: Analysis of spatial gradients in lung cancer mortality rates in north carolina counties. *Biometrics*, 37(2):259–269, 1981.

B. H. Margolin, B. S. Kim, and K. J. Risko. The ames salmonella/microsome mutagenicity assay: Issues of inference and validation. *Journal of the American Statistical Association*, 84(407):651–661, 1989.

P. McCullagh, and J. A. Nelder. *Generalized Linear Models (2nd Edition)*. Chapman & Hall, 1989.

C. E. McCulloch, and S. R. Searle. *Generalized, Linear and Mixed Models*. John Wiley & Sons, New York, USA, 2001.

G. J. McLachlan. On the em algorithm for overdispersed count data. *Statistical Methods in Medical Research*, 6(1):76–98, 1997.

R. Mian, and S. Paul. Estimation for zero-inflated over-dispersed count data model with missing response. *Statistics in medicine*, 35(30):5603–5624, 2016.

A. J. Miller. *Subset Selection in Regression*. London: Chapman and Hall, 1990.

G. Molenberghs, and G. Verbeke. *Models for Discrete Longitudinal Data*. Sprnger, New York, 2005.

A. Muñoz, V. Carey, J. P. Schouten, M. Segal, and B. Rosner. A parametric family of correlation structures for the analysis of longitudinal data. *Biometrics*, 48: 733–742, 1992.

A. Munoz, B. Rosner, and V. Carey. Regression analysis in the presence of heterogeneous intraclass correlations. *Biometrics*, 42(3):653–658, 1986.

J. A. Nelder, and D. Pregibon. An extended quasi-likelihood function. *Biometrika*, 74:221–232, 1987.

J. M. Neuhaus, and J. D. Kalbfleish. Between- and within-cluster covariate effects in the analysis of clustered data. *Biometrics*, 54:638–645, 1998.

J. Neyman. Optimal asymptotic tests of composite statistical hypotheses. *Probability and statistics*, 57:213–234, 1959.

V. Núñez Anton, and G. G. Woodworth. Analysis of longitudinal data with unequally spaced observations and time-dependent correlated errors. *Biometrics*, 50:445–456, 1994.

R. J. O'Hara-Hines. Comparison of two covariance structures in the analysis of clusterd polytomous data using generalized estimating equations. *Biometrics*, 54:312–316, 1998.

A. B. Owen. *Empirical likelihood*. New York: Chapman and Hall, CRC, 2001.

M. C. Paik. The generalized estimating equation approach when data are not missing completely at random. *Journal of the American Statistical Association*, 92:1320–1329, 1997.

W. Pan. On the robust variance estimator in generalised estimating equations. *Biomtrika*, 88:901–906, 2001a.

W. Pan. Akaike's information criterion in generalized estimating equations. *Biometrics*, 57:120–125, 2001b.

H. D. Patterson, and R. Thompson. Recovery of inter-block information when block sizes are unequal. *Biometrika*, 8:545–554, 1971.

S. R. Paul. Maximum likelihood estimation of intraclass correlation in the analysis of familial data: Estimating equation approach. *Biometrika*, 77(3):549–555, 1990.

S. R. Paul. Quadratic estimating equations for the estimation of regression and dispersion parameters in the analysis of proportions. *Sankhya B*, 63:43–55, 2001.

S. R. Paul, and T. Banerjee. Analysis of two-way layout of count data involving multiple counts in each cell. *Journal of the American Statistical Association*, 93 (444):1419–1429, 1998.

S. R. Paul, and R. L. Plackett. Inference sensitivity for poisson mixtures. *Biometrika*, 65(3):591–602, 1978.

S. R. Paul. Analysis of proportions of affected foetuses in teratological experiments. *Biometrics*, 38:361–370, 1982.

S. R. Paul, and A. S. Islam. Analysis of proportions based on parametric and semi-parametric models. *Biometrics*, 51:1400–1410, 1995.

M. S. Pepe, and G. L. Anderson. A cautionary note on inference for marginal regression models with longitudinal data and general correlated response data. *Communications in Statistics, Part B - Simulation and Computation*, 23:939–951, 1994.

W. W. Piegorsch. Maximum likelihood estimation for the negative binomial dispersion parameter. *Biometrics*, 46(3):863–867, 1990.

J. C. Pinheiro, and D. M. Bates. *Mixed-Effects Models in S and S-PLUS*. Springer-Verlag, New York, 2000.

J. S. Preisser, K. K. Lohman, and P. J. Rathouz. Performance of weighted estimating equations for longitudinal binary data with drop-outs missing at random. *Statistics in Medicine*, 21:3035–3054, 2002.

R. L. Prentice, and L. P. Zhao. Estimating equations for parameters in means and covariances of multivariate discrete and continuous responses. *Biometrics*, 47: 825–839, 1991.

C. J. Price, C. A. Kimmel, J. D. George, and M. C. Marr. The developmental toxicity of ethylene glycol in mice. *Fundamental and Applied Toxicology*, 81:113–127, 1985.

J. Qin, and J. Lawless. Empirical likelihood and generalized estimating equations. *The Annals of Statistics*, 22:300–325, 1994.

A. Qu, B. G. Lindsay, and Bing Li. Improving generalised estimating equations using quadratic inference functions. *Biometrik*, 87:823–836, 2000.

A. Qu, and P. X. K. Song. Assessing robustness of generalised estimating equations and quadratic inference functions. *Biometrika*, 91(2):447–459, 2004.

A. Qu, J. J. Lee, and B. G. Lindsay. Model diagnostic tests for selecting informative correlation structure in correlated data. *Biometrika*, 95(4):891–905, 12 2008.

A. E. Raftery, D. Madigan, and J. A. Hoeting. Bayesian model averaging for linear regression models. *Journal of the American Statistical Association*, 92(437): 179–191, 1997.

J. N. K. Ran, and I. Molina. *Small Area Estimation,*, 2nd ed. Wiley, New York, 2015.

C. R. Rao. *Linear Statistical Inference and its Applications*. Wiley, 1965.

P. J. Rathouz, and L. Gao. Generalized linear models with unspecified reference distribution. *Biostatistics*, 10:205–218, 2009.

J. M. Robin, A. Rotnitzky, and L. P. Zhao. Analysis of semiparametric regression models for repeated outcomes in the presence of missing data. *Journal of the American Statistical Association*, 90:106–121, 1995.

J. M. Robins, and A. Rotnitzky. Semiparametric efficiency in multivariate regression models with missing data. *Journal of the American Statistical Association*, 90 (429):122–129, 1995.

B. Rosner. Multivariate methods in ophthalmology with application to other paired-data situations (c/r: V46 p523-531). *Biometrics*, 40:1025–1035, 1984.

G. J. S. Ross, and D. A. Preece. The negative binomial distribution. *Journal of the Royal Statistical Society: Series D (The Statistician)*, 34(3):323–335, 1985.

A. Rotnitzky, and N. P. Jewell. Hypothesis testing of regression parameters in semi-parametric generalized linear models for cluster correlated data. *Biometrika*, 77: 485–497, 1990.

P. J. Rousseeuw, and B. C. van Zomeren. Unmasking multivariate outliers and leverage points. *Journal of the American Statistical Association*, 85:633–639, 1990.

D. B. Rubin. Inference and missing data. *Biometrika*, 63:581–592, 1976.

S. Galbraith, J. A. Daniel, and B. Vissel. A study of clustered data and approaches to its analysis. *Journal of Neuroscience*, 30(32):10601–10608, 2010.

K. Saha, and S. Paul. Bias-corrected maximum likelihood estimator of the negative binomial dispersion parameter. *Biometrics*, 61(1):179–185, 2005.

H. Sahai, A. Khuri, and C. H. Kapadia. A second bibliography on variance components. *Communications in Statistics-theory and Methods*, 14:63–115, 1985.

H. Sahai. A bibliography on variance components. *International Statistical Review/Revue Internationale de Statistique*, 47(2):177–222, 1979.

S. K. Sahu, and G. O. Roberts. On convergence of the em algorithmand the gibbs sampler. *Statistics and Computing*, 9(1):55–64, 1999.

D. E. W. Schumann, and R. A. Bradley. The comparison of the sensitivities of similar experiments: Theory. *The Annals of Mathematical Statistics*, 13(4):902–920, 1957.

S. R. Searle, G. Casella, and C. E. McCulloch. *Variance Components*, volume 391. John Wiley & Sons, 2009.

P. E. Shrout, and J. L. Fleiss. Intraclass correlations: uses in assessing rater reliability. *Psychological Bulletin*, 86(2):420, 1979.

J. Shults, and R. N. Chaganty. Analysis of serially correlated data using quasi-likelihood squares. *Biometrics*, 54:1622–1630, 1998.

J. Shults, and J. M. Hilbe. *Quasi-Least Squares Regression*. Chapman & Hall/CRC Monographs on Statistics & Applied Probability. Taylor & Francis, 2014.

S. Sinha, and T. Maiti. Analysis of matched case-control data in presence of nonignorable missing exposure. *Biometrics*, 64(1):106–114, 2008.

A. Sklar. Fonctions de répartition à n de dimensions et leursmarges. *Paris: Publications de l'Institut de Statistique de l'Université de Paris 8*, 143:229–231, 1959.

P. X.-K. Song. Correlated data analysis. modeling. *Analytics and Applications*. *Springer, New York*, 2007.

P. X.-K. Song, Z. C. Jiang, E. Park, and A. Qu. Quadratic inference functions in marginal models for longitudinal data. *Statistics in Medicine*, 28(29):3683–3696, 2009.

M. F. Sowers, M. Crutchfield, J. F. Randolph, B. Shapiro, B. Zhang, M. L. Pietra, and M. A. Schork. Urinary ovarian and gonadotrophin hormone levels in premenopausal women with low bone mass. *Journal of Bone Mining Research*, 13:1191–1202, 1998.

P. T. Spellman, G. Sherlock, M. Q. Zhang, et al. Comprehensive identification of cell cycle-regulated genes of the yeast saccharomyces cerevisiae by microarray hybridization. *Molecular Biology of Cell*, 9:3273–3297, 1998.

B. C. Sutradhar, and K. Das. On the efficiency of regression estimators in generalised linear models for longitudinal data. *Biometrika*, 86(2):459–465, 1999.

P. F. Thall, and S. C. Vail. Some covariance models for longitudinal count data with overdispersion. *Biometrics*, 46:657–671, 1990.

W. A. Thompson Jr. The problem of negative estimates of variance components. *The Annals of Mathematical Statistics*, 33:273–289, 1962.

R. J. Tibshirani. Regression shrinkage and selection via the lasso. *Journal of the Royal Statistical Society, Ser. B*, 58:267–288, 1996.

P. Tishler, A. Donner, J. O. Taylor, and E. H. Kass. Familial aggregation of blood pressure in very young children. *CVD Epidemiology Newsletter*, 22(45), 1977.

D. Trègouët, P. Ducimetière, and L. Tiret. Testing association between candidate-gene markers and phenotype in related individuals, by use of estimating equations. *American Journal of Human Genetics*, 61:189–199, 1997.

A. B. Troxel, S. L. Lipsitz, and T. A. Brennan. Weighted estimating equations with nonignorably missing response data. *Biometrics*, 53:857–869, 1997.

R. S. Tsay. Regression models with time series errors. *Journal of the American Statistical Association*, 79(385):118–124, 1984.

M. C. K. Tweedie. An index which distinguishes between some important exponential families. In *Statistics: Applications and new directions: Proc. Indian Statistical Institute Golden Jubilee International Conference*, volume 579, 579–604, 1984.

G. Verbeke. *Models for Discrete Longitudinal Data. Springer Series in Statistics.* Springer, 2005.

L. Wang, J. H. Zhou, and A. Qu. Penalized generalized estimating equations for high-dimensional longitudinal data analysis. *Biometrics*, 68(1):353–360, 2012a.

L. Wang, and A. Qu. Consistent model selection and data-driven smooth tests for longitudinal data in the estimating equations approach. *Journal of the Royal Statistical Society: Series B*, 71(1):177–190, 2009.

P. Wang, G-F. Tsai, and A. Qu. Conditional inference functions for mixed-effects models with unspecified random-effects distribution. *Journal of the American Statistical Association*, 107(498):725–736, 2012b.

Y-G. Wang. A quasi-likelihood approach for ordered categorical data with overdispersion. *Biometrics*, 52(4):1252–1258, 1996.

Y-G. Wang. Estimating equations for removal data analysis. *Biometrics*, 55(4): 1263–1268, 1999a.

Y-G. Wang. Estimating equations with nonignorably missing response data. *Biometrics*, 55(3):984–989, 1999b.

Y-G. Wang, and V. J. Carey. Working correlation structure misspecification, estimation and covariate design: Implications for generalised estimating equations performance. *Biometrika*, 90(1):29–41, 2003.

Y-G. Wang, and V. J. Carey. Unbiased estimating equations from working correlation models for irregularly timed repeated measures. *Journal of the American Statistical Association*, 99(467):845–853, 2004.

Y-G. Wang, and L-Y. Hin. Modeling strategies in longitudinal data analysis: Covariate,variance function and correlations structure selection. *Computational Statistics and Data Analysis*, 54(5):3359–3370, 2009.

Y-G. Wang, and X. Lin. Effects of variance-function misspecification in analysis of longitudinal data. *Biometrics*, 61(2):413–421, 2005.

Y-G. Wang, and Y. D. Zhao. Weighted rank regression for clustered data analysis. *Biometrics*, 64:39–45, 2008.

Y-G. Wang, and Y. N. Zhao. A modified pseudolikelihood approach for analysis of longitudinal data. *Biometrics*, 63(3):681–689, 2007.

Y-G. Wang, and M. Zhu. Rank-based regression for analysis of repeated measures. *Biometrika*, 93:459–464, 2006.

Y-G. Wang, X. Lin, and M. Zhu. Robust estimating functions and bias correction for longitudinal data analysis. *Biometrics*, 61(3):684–691, 2005.

Y-G. Wang, Q. Shao, and M. Zhu. Quantile regression without the curse of unsmoothness. *Computational Statistics and Data Analysis*, 52:3696–3705, 2009.

J. H. Ware, S. Lipsitz, and F. E. Speizer. Issue in the analysis of repeated categorical outcomes. *Statistics in Medicine*, 7:95–107, 1988.

R. W. M. Wedderburn. Quasi-likelihood functions, generalized linear models and the gauss-newton method. *Biometrika*, 61:439–47, 1974.

L. J. Wei, and J. M. Lachin. Two-sample asymptotically distribution-free tests for incomplete multivariate observations. *Journal of the American Statistical Association*, 79(387):653–661, 1984.

C. S. Weil. Selection of the valid number of sampling units and a consideration of their combination in toxicological studies involving reproduction, teratogenesis or carcinogenesis. *Food and cosmetics toxicology*, 8(2):177–182, 1970.

M. Cho, R. E. Weiss, and M. Yanuzzi. On bayesian calculations for mixture priors and likelihoods. *Statistics in Medicine*, 18:1555–1570, 1999.

H. White. Maximum likelihood estimation of misspecified models. *Econometrica*, 50(1):1–25, 1982.

P. Whittle. Gaussian estimation in stationary time series. *Bulletin of the International Statistical Institute*, 39:1–26, 1961.

D. A. Williams. Dose-response models for teratological experiments. *Biometrics*, 43:1013–1016, 1987.

J. M. Williamson, S. Datta, and G. A. Satten. Marginal analyses of clustered data when cluster size is informative. *Biometrics*, 59(1):36–42, 2003.

R. F. Woolson, and W. R. Clarke. Analysis of categorical incomplete longitudinal data. *Journal of the Royal Statistical Society. Series A (General)*, 147(1):87–99, 1984.

H. Wu, and A. A. Ding. Population hiv-1 dynamics in vivo: applicable models and inferential tools for virological data from aids clinical trials. *Biometrics*, 55: 410–418, 1999.

P. R. Xu, L. X. Zhu, and Y. Li. Ultrahigh dimensional time course feature selection. *Biometrics*, 70:356—365, 2014.

G. Yin, and J. Cai. Quantile regression models with multivariate failure time data. *Biometrics*, 61:151–161, 2005.

S. L. Zeger, and K-Y. Liang. Longitudinal data analysis for discrete and continuous outcomes. *Biometrics*, 42(1):121–130, 1986.

S. L. Zeger, and B. Qaqish. Markov regression models for time series: A quasi-likelihood approach. *Biometrics*, 44:1019–1031, 1988.

L. Zeng, and R. J. Cook. Transition models for multivariate longitudinal binary data. *Journal of the American Statistical Association*, 102(477):211–223, 2007.

C-H. Zhang, and J. Huang. The sparsity and bias of the lasso selection in high-dimensional linear regression. *The Annals of Statistics*, 36(4):1567–1594, 2008.

D. Zhang, X. H. Lin, J. Raz, and M. F. Sowers. Semiparametric stochastic mixed models for longitudinal data. *Journal of the American Statistical Association*, 93:710–719, 1998.

W. P. Zhang, C. L. Leng, and C. Y. Tang. A joint modelling approach for longitudinal studies. *Journal of the Royal Statistical Society: Series B*, 77(1):219–238, 2015.

X. M. Zhang, and S. Paul. Modified gaussian estimation for correlated binary data. *Biometrical Journal*, 55(6):885–898, 2013.

L. P. Zhao, and R. L. Prentice. Correlated binary regression using a quadratic expo-nential model. *Biometrika*, 77:642–648, 1990.

L. P. Zhu, K. Xu, R. Li, and W. Zhong. Projection correlation between two random vectors. *Biometrika*, 104:829–843, 2017.

A. Ziegler. *Generalized Estimating Equations (Lecture Notes in Statistics)*. Pub-lisher: Springer, 2011.

H. Zou. The adaptive lasso and its oracle properties. *Journal of the American Statis-tical Association*, 476:1418–1429, 2006.

H. Zou, and Hastie T. Regularization and variable selection via the elastic net. *Journal of the Royal Statistical Society, Ser. B*, 67:301–320, 2005.

S. J. Zyzanski, S. A. Flocke, and L. M. Dickinson. On the nature and analysis of clustered data. *The Annals of Family Medicine*, 2(3):199–200, 2004.

Author Index

Subject Index